Lessons From Fukushima

Yuko Fujigaki
Editor

Lessons From Fukushima

Japanese Case Studies on Science,
Technology and Society

 Springer

Editor
Yuko Fujigaki
Graduate School of Arts and Sciences
University of Tokyo
Tokyo, Tokyo
Japan

ISBN 978-3-319-15352-0   ISBN 978-3-319-15353-7 (eBook)
DOI 10.1007/978-3-319-15353-7

Library of Congress Control Number: 2015931539

Springer Cham Heidelberg New York Dordrecht London
© Springer International Publishing Switzerland 2015
This work is subject to copyright. All rights are reserved by the Publisher, whether the whole or part of the material is concerned, specifically the rights of translation, reprinting, reuse of illustrations, recitation, broadcasting, reproduction on microfilms or in any other physical way, and transmission or information storage and retrieval, electronic adaptation, computer software, or by similar or dissimilar methodology now known or hereafter developed.
The use of general descriptive names, registered names, trademarks, service marks, etc. in this publication does not imply, even in the absence of a specific statement, that such names are exempt from the relevant protective laws and regulations and therefore free for general use.
The publisher, the authors and the editors are safe to assume that the advice and information in this book are believed to be true and accurate at the date of publication. Neither the publisher nor the authors or the editors give a warranty, express or implied, with respect to the material contained herein or for any errors or omissions that may have been made.

Printed on acid-free paper

Springer International Publishing AG Switzerland is part of Springer Science+Business Media (www.springer.com)

Foreword by Rethy Chhem

The destruction brought down and across the Japanese coast in March 2011 has returned to us our fear and our uncertainty. We had assumed mastery where misunderstanding and negligence lay. Above all, we have seen the world, and ourselves inside it, brought to cold reflections:

How have our systems failed?
How have we failed to prevent?
How have we failed to imagine?

The third charge, our most severe and accusatory, encompasses the whole of our missteps and malpractice towards this disaster and all before it: We may have failed to imagine.

Once, requiring only a crude understanding of possibilities and outcomes, wielding technologies of risk was a seemingly simple affair. Their behaviour we assumed to be known, controlled and local. We managed the benefits and risks of these instruments with predictive laws, and thus, predictive remediation when we experience failure. The uncertainties of our technology and its scientific basis were to be uncovered through progress, never beyond the limits of investigation and never included as a feature of science.

Now, in spite of this, our increasingly complex social world has challenged us to revise our entrenched narrowness of problem solving. We witness physical destruction pushing far beyond the normative framework of scientific understanding and must draw new ways of looking, from a variety of fields, to these intricate problems. We must filter our vision through the pages of new volumes.

In sum, we must reimagine.

With over 30 years' experience in radiation medicine as a practicing radiologist, a historian of science and social scientist—along with numerous publications on clinical radiology, radiation protection and a Science-Technology-Society (STS) handbook for Fukushima Medical University (FMU)—I have developed a unique perspective on this emergent requirement of social understanding in disaster relief. The interactions between science, technology and society are immensely important to gaining a clear analysis of disaster towards remediation. With such vision, we can uncover the complex and wondrous assemblage of the nuclear reactor beneath the mechanistic surface: A singular nexus of the social, the political and the material, powered by courage and ambition. Likewise, we can investigate its failures with similar acuity.

When the compound disaster of 3/11 occurred, I witnessed efforts to mitigate the short and mid-term damage of the disaster troubled by immense difficulties. As a Visiting Professor to Nagasaki and Hiroshima Universities as well as FMU, where I was teaching STS to healthcare professionals during the crisis and the recovery phase, I became acutely aware of the limitations of these emergency efforts. These efforts—which, particularly at the center of disaster, were often heroic—were not simply afflicted by the conditions of radioactive fallout, but also by misunderstanding of the unique human afflictions brought on by its after effects. Moreover, the communication of risk to the public remained greatly hindered by this lack of comprehension surrounding the social issues of the disaster. It seemed to us that otherwise clean, easily assuaged problems had lifted from the textbook and into the social, taking on the form of acute psychological illness. In other words, information that performed neatly in theory would complicate and disperse in this very alive, very human situation.

This social element proved critical, with few of those professionals leading the emergency response possessing an expert understanding. The mass migration of Japanese citizens from the disaster area to safer ground, the dislocated youth, adults and elderly: all were shuttled away from their realities to form temporary lives for indefinite time. Lives caught in jeopardy, journeying between the irradiated shoreline to the anxieties of unemployment and social exclusion.

Had the capacity for exploring the possibilities of our technology led to an inability to control its development? Had our dependency on the existence of this technology simply blinded us to their complex dangers? The triple disaster of 3/11 had thoroughly pierced the rigidity of disaster preparedness. To reimagine and reform, we must spread the boundaries of our understanding.

The publication you hold is a welcome response to this challenge. By first addressing major questions related to the social impacts of the Fukushima Disaster, Professor Yuko Fujigaki continues a potent argument towards broadening our understanding of the placement of the Fukushima disaster in the mutable folds of society.

Particularly important to our understanding, the author illustrates the historical bearing of nuclear technology—from the atomic bombings of late 1945 to the chief concern of Fukushima—on the island nation and how these technologies would become interwoven within the complex national character of Japan. This historical perspective provides Professor Fujigaki with the grounds for crucial comparisons with other diseases (Minimata and Itai-Itai disease) and with other cases (HIV-tainted-blood scandal and Winny case) in part two of publication in order to investigate the complex relationship between science, technology and society. This series of similar yet distinct technological concerns allows the author to trace the actors and assemblages that construct the foundations by which technologies are envisioned, enacted and remedied, particular once they fail to serve us correctly.

Professor Fujigaki's breadth of knowledge in the field of STS and her leadership in the Japanese STS community brilliantly showcases individual efforts to progress our knowledge of risk communication, the complex layers of interactions within society, and scientific and technological activity. The creation of this publication itself is evidence of the considerable efforts shown by the professor, having led the process of gathering Japanese experts to contribute to this uniquely valuable publication. Through such academic ambitions, we continue to push outward and expand the frame of our perception to the convoluted realities of disaster, particularly where our social fabric has frayed most conspicuously, where lives have been troubled most deeply.

And with a new vision atop these pillars of experience, relationships between particles and people attain equal importance, each requiring sound and balanced analysis. And with this analysis, we can survey our errors at its truest extent, and bring science to greater utility.

Dr. Rethy Chhem
Visiting Professor
Fukushima Medical University
Hiroshima University
Nagasaki University

Foreword by Wiebe E. Bijker

This volume provides a rich collection of studies to help us understand the "Great East Japan Earthquake", or rather the triple disaster—of earthquake, tsunami and nuclear power plant failure—that happened in and around Fukushima in March 2011. The chapters address questions about the historical development of the Japanese nuclear program, risk management and communication, new forms of participatory technology planning, pollution and public health, and the governance of technology. These questions are addressed by using a broad range of approaches from the interdisciplinary spectrum that makes up the field of Science, Technology and Society studies (STS).

But it does more than addressing these important questions generated by "Fukushima". The volume, through its lens of the triple disaster, creates a fascinating cross-section of Japanese society. Using STS perspectives, Yuko Fujigaki and her colleagues sketch a picture of Japanese society that is at the same time familiar to my European eyes, and fascinatingly different.

The first time I visited Japan in 2003, this was—quite ironically in the context of this volume—on invitation by the Central Research Institute of Electric Power Industry (CRIEPI) and the Japan Science and Technology Agency (JST), to discuss new forms of citizens' participation in locating power stations. Ever since, I have been struck by the potential richness of conversations between East and West, between Japan and Europe/US. The founding of the Japanese Society for Science and Technology Studies (JSSTS) in 2001 and the successful large international conference that JSSTS organized with the Society for Social Studies of Science (4S) in Tokyo in 2010 have underlined how fruitful such STS-inspired comparisons and conversations can be.

It is a triviality to note that societies differ, also in their socio-technical make-up. More interesting is the question whether STS, as an emerging discipline that studies such societies as constituted by science and technology, differs between the various subcontinents of our globe too. Readers are invited to explore in this volume whether there is something as a Japanese style of doing STS. The volume certainly provides an illuminating view of Japanese society.

In the case of this volume moreover, the STS authors do more than study Japanese society and the triple disaster. They engage with the burning societal questions that were generated by the Great East Japan Earthquake. Japanese STS researchers have reacted to the Fukushima events by joining government advisory committees, doing post-disaster communication, conducting policy experiments such as deliberative polling, participating in public discussions and providing historical and critical analyses of what the Japanese citizens saw happening around them. And finally they intensively contributed with STS colleagues from abroad to help the IAEA and Fukushima Medical University to develop an STS-inspired curriculum to better prepare medical doctors for events such as the triple disaster.

This book, in sum, not only provides an enlightening analysis of the Great East Japan Earthquake, but also offers an intriguing view of Japanese techno-scientific culture, and gives an inspiring example of how STS scholars could engage with the world around them.

Wiebe E. Bijker
Professor of Technology and Society
Maastricht University

Past-President
Society for Social Studies of Science (4S)

Preface

The Fukushima nuclear power plant accidents after an earthquake in March 2011 have drawn the world's attention to the relationship between science, technology and society in the high-technology setting of Japan. For example, how are the nuclear power plants embedded in political, economic and social contexts in Japan? Under what kinds of relationships between science, technology and society are such accidents produced? In addition, how are these relationships constructed historically? This book provides a case analysis on the Triple Disaster (i.e. earthquake, tsunami and nuclear power plants) to address the first two questions and also provides analysis on Minamata disease (Mercury pollution) and Itai-Itai disease (Cadmium pollution) to examine the last question.

The first question is one that I received from Ulrike Felt, Professor at the University of Vienna, just after the earthquake in April 2011. Professor Felt posed this question at the "STS 20+20" (Science and Technology Studies: The Next Twenty: A conference reflecting on the past 20 years of STS graduate study and looking ahead to the next 20) at Harvard University. This is a very important question for analysing the Fukushima accidents from the perspectives from Science and Technology Studies. Since the conference, we Japanese researchers have eagerly examined this first question as well as the second question; for example, I chaired and played an discussant in the joint plenary of the History of Science Society (HSS), the Society for History of Science (SHOT), and the Society for Social Studies of Science (4S) in Cleveland on November 3, 2011, on "Dealing with Disasters: Perspectives on Fukushima from the History and Social Studies of Science and Technology". Discussions have also been developed in the sessions at the 4S and the European Association of Science and Technology Studies (EASST) joint conference in October 2012 in Copenhagen. We make full use of these discussions in four chapters in Part I.

To answer the last question, we deal with several case analyses in Part II. These analyses are based on the Japanese STS textbook edited in 2005 in Japanese as a result of a project funded by the Japan Science and Technology Agency. The project began in January 2002, and we received helpful advice from Sheila Jasanoff, a Professor at Harvard University, at the Science and Democracy Meeting held in

Berlin in 2002. Based on her advice, we held an international workshop in December 2003. Michel Callon, Brian Wynne, Ulrike Felt, Rob Hagendijk and Thomas Gieryn gave us much useful feedback. In the 4S/EASST 2004 Paris meeting, we organised a session on results of the project, in collaboration with Michel Callon and Edward Hackett, who were commentators. After we submitted a final report to Japan Science and Technology Agency (JST), we published the STS textbook, *Case Analysis and Theoretical Concepts for Science and Technology Studies* (University of Tokyo Press 2005). Part II of this book is a selected, revised version of this textbook.

In the process of planning to publish this book, the editor received insightful comments from Rethy Chhem. I would like to express my gratitude to Rethy for the realization of this book.

Yuko Fujigaki
Professor, The University of Tokyo

Contents

1 Introduction 1
  Yuko Fujigaki

Part I Lessons from Fukushima

2 The Processes Through Which Nuclear Power Plants
  Are Embedded in Political, Economic, and Social
  Contexts in Japan... 7
  Yuko Fujigaki

3 Agenda Building Intervention of Socio-Scientific Issues:
  A Science Media Centre of Japan Perspective 27
  Mikihito Tanaka

4 Rhetorical Marginalization of Science and Democracy:
  Politics in Risk Discourse on Radioactive Risks in Japan 57
  Hideyuki Hirakawa and Masashi Shirabe

5 Public Participation in Decision-Making on Energy Policy:
  The Case of the "National Discussion" After the Fukushima
  Accident.. 87
  Naoyuki Mikami

**Part II Historical Construction of Science, Technology
           and Society Relationship**

6 Minamata Disease: Interaction Between Government,
  Scientists, and Media 125
  Shigeo Sugiyama

7	Itai-itai Disease: Lessons for the Way to Environmental **Regeneration** 141
	Masanori Kaji

8	**The Monju Trial: Nuclear Controversy in Japan** 167
	Tadashi Kobayashi and Minako Kusafuka

9	AIDS Patients Due to Transfusion of HIV Infected, **Non-heat-treated Blood Products**........................ 195
	Yoshiyuki Hirono

10	Winny Criminal Case: How Have Controversial Science, Technology, and Society Problems Been *Solved* While **Avoiding Conflicts?**................................. 219
	Masahi Shirabe

Index 239

Chapter 1
Introduction

Yuko Fujigaki

This book consists of two parts. Part I provides case analyses on the Triple Disaster (i.e., earthquake, tsunami, and nuclear power plants) in Japan to seek answers to two questions: first, how are nuclear power plants embedded in political, economic, and social contexts in Japan? Second, under what kinds of relationships between science, technology, and society are such accidents produced? Chapter 2 reflects Japanese history after the atomic bomb was dropped in Hiroshima and Nagasaki in 1945, followed by the passing of the budget plan for nuclear power in the Japanese Diet in 1954, based on the U.S. presidential address on "Atoms for Peace" by Eisenhower in 1953. This chapter tries to describe how nuclear power plants were embedded in political, economic, and social contexts in Japan from 1950 to 1970, during which period most power plants were designed to be constructed. In addition, this chapter analyzes the communication disaster after Triple Disaster in Japan. Through analysis on the communication disaster, the author describes the relationships between science, technology, and society in Japan, under which such accidents were produced.

Chapter 3 describes risk-communication after the accidents based on analysis of social media in Japan. There was a discrepancy between what was conveyed by mass media and by social media. At large, mass media had a tendency to convey the "safety" information, whereas social media conveyed not only safety information, but also information on risk. The author managed the website "Science Media Centre" on the web after the accident and summarizes his experiences with conducting this site.

Chapter 4 deals with politics in the risk-discourse on the radioactive risk in Japan. This paper was also presented at the Science and Democracy Network in Paris in 2012. The authors have also published a Japanese book on this topic. This chapter is the grand sum of these activities.

Y. Fujigaki (✉)
The University of Tokyo, Tokyo, Japan
e-mail: fujigaki@idea.c.u-tokyo.ac.jp

Following the disaster, Japanese society engaged in serious discussions on future energy sources for an industrial society. Chapter 5 introduces the 2012 "Deliberative Poll" survey on future energy in Japan. The author was one of the conductors of this survey, and summarizes the experience of this process in Japanese society.

Part II provides case analyses on Minamata disease (Mercury pollution), Itai-Itai Disease (Cadmium pollution), controversy on nuclear power plants, drug-induced disease (HIV-infected blood products), and so on to answer the question of how these relationships among science, technology, and society are constructed historically. Chapter 6 deals with Minamata disease. The first patient suffering from what became known as Minamata disease was identified in 1956; later, the patient's condition was found to be caused by alkyl-mercury poisoning from polluted water. However, the process of analyzing the cause of the disease was not so straightforward. Science is always "in the making"; therefore, if the government does not have the countermeasure until the causes of the disease is proved precisely, then delay of the countermeasure leads to a substantial increase in the number of patients.

On the contrary, in the case of Itai-Itai disease, which is presented in Chap. 7, countermeasures to deal with the wastewater occurred much faster than in the case of Minamata disease. The first head of the Department of Public Environmental Hazard showed the "precautionary principle," stating that "although there was insufficient evidence to determine that cadmium was as the cause of the disease, there was not enough time to wait until the evidence was completely unveiled—it was time to take action against the disease. We will continue the scientific investigations." In addition, the Residents Association and the Mitsui Company concluded a Pollution Control Agreement in 1972, and this agreement enables the residents to conduct wastewater inspections in collaboration with attorneys, experts, and citizens from other districts. This activity succeeded in reducing the levels of cadmium in wastewater. Thus, Itai-Itai disease is an outstanding case in which Japanese society succeeded in utilizing the "precautionary principle" in environmental problems. However, this kind of precautionary principle and experience of public involvement has not been utilized in the field of nuclear power.

Chapter 8 reveals what happened in citizen movement in the field of nuclear power. It deals with trials regarding the fast-breeder-reactor Monju. In May 1983, the Japanese government (Prime Minister) approved the establishment of Monju Nuclear Power Plant in the Tsuruga district, Fukui Prefecture. However, an incident involving the leakage of sodium at the Monju plant occurred in December 1995, and the plant was required to stop operations. This chapter analyzes the consequences of the legal action by local residents against the government including the High Court decision in 2003 and the Supreme Court decision in 2005. Legal experts pointed to the lack of public participation in the administrative process in the initial approval of the construction of the plant. They said that administrative discretion was originally a tool for step-by-step handling of complicated problems with the involvement of stakeholders. However, this kind of administrative discretion did not function in the decision-making process for the establishment of Monju. These facts indicate that, in the field of nuclear power management, public participation was insufficient and Japanese society failed to construct a "public sphere" to discuss the validity of construction of nuclear power plants as well as to discuss the future of energy in the 21st century.

Chapter 9 presents questions regarding the delay of countermeasures by the government. It provides analysis on several trials by patients who contracted AIDS due to transfusions of HIV-infected, non-heated blood products in the 1980s and 1990s. The doctor who was responsible for the treatment was found not guilty by the decision of the court based on the "new theory of negligence". However, if the doctor was not accountable, who is responsible for the delay of countermeasures against the transfusion of HIV-infected, non-heated blood products? It invites questions on science policy. When the Bovine Spongiform Encephalopathy (BSE) scandal in Japan was criticized in the journal Nature in 2000, one article stated that "the fears are all the more palpable because the Japanese government has shown that it cannot be trusted to do the right thing in such public-health cases" (Nature, vol. 413 P. 333, 27 September 2001). It also indicates that a "Lack of scientific evidence was also the excuse made for not taking action in the earlier cases". Thus, this case shows the tendency for the Japanese government to be reluctant to adopt the precautionary principle in most areas of public health, even though they used it successfully in the Itai-Itai disease case.

Chapter 10 provides analysis on advanced technology and law by examining the Winny Scandal. The information professional who developed the "Winny" software (a file sharing system through Pear to Pear (P2P) system) was arrested in 2002 but ultimately found not guilty after three courts decisions. This chapter deals with the controversy on advanced technology which has a high possibility of copyright infringement.

Based on these chapters, we can summarize the relationships between science, technology, and society that have been constructed historically in Japan as follows. (1) There existed a "precautionary principle" in the 1970s in the case of environmental issues; however, in the field of nuclear power plants, this principle did not work. (2) From the observation of administrative lawsuits, several experts pointed to the lack of public engagement in the administrative process in the initial approval of the construction of the plants. These points, which came to light after the controversy in the administrative lawsuit, were not utilized for nuclear power safety discussion after the lawsuit. (3) New relationships between science, technology, and society are now being constructed in the fields of new technology, e.g., information technology; however, this has had little effect on the historically rigidly constructed relationships among these spheres in nuclear power energy.

Part I
Lessons from Fukushima

Chapter 2
The Processes Through Which Nuclear Power Plants Are Embedded in Political, Economic, and Social Contexts in Japan

Yuko Fujigaki

Abstract To analyze the process through which nuclear power plants are embedded in political, economic, and social contexts in Japan, this chapter first deals with a brief history on nuclear power plants in Japan and explore cultural acceptance of nuclear energy, the role of nuclear energy in the political system, and the status of the nuclear industry. Then I will examine the politics of "unexpected" or "beyond expectation" discourse using reports by the National Diet, by the Cabinet and by Independent Investigation Commission to survey the source of legitimate expertise in this domain. Furthermore, this paper deals with the communication disaster after the accidents as well as public debate in Japanese society to analyze the role of media and the culture of public debate over complex techno-scientific issues. From these analyses, we can determine that segregation was established between sites that accepted nuclear power plants before the 1970s and sites without nuclear power plants. After the accidents of March 11, 2011, this segregation expanded between these sites as well as within each site. In addition, discussions about whether to consider the accidents as universal lessons from Fukushima or to regard the accidents as culturally specific leads us to a discussion on technological culture with relevance to techno-orientalism.

2.1 Introduction

How are nuclear power plants embedded in political, economic, and social contexts in Japan? To answer this question, we should consider the local techno-scientific-political culture in Japan. In the survey of techno-scientific-political culture, Felt (2013a, b) highlighted the importance of long-term, comparative research to reflect on: (A) cultural acceptance of nuclear energy, (B) the role of nuclear energy in the political system, (C) the status of the nuclear industry, (D) the source of legitimate

Y. Fujigaki (✉)
The University of Tokyo, Tokyo, Japan
e-mail: fujigaki@idea.c.u-tokyo.ac.jp

expertise in this domain, (E) the role of media, and (F) the culture of public debate over complex techno-scientific issues.

In the following sections, I will examine these six items. In Sect. 2.2, I will present a brief history on nuclear power plants in Japan and explore items (A), (B), and (C) through this historical analysis. In Sect. 2.3, I will examine the politics of "unexpected" or "beyond expectation" discourse using a report by the National Diet (2012), a report by the Cabinet (2012), and a report by Independent Investigation Commission on the Fukushima Daiichi Nuclear Accident (2012). Item (D) will be surveyed in this section. Section 2.4 will deal with the communication disaster after the accidents as well as public debate in Japanese society. Through the analysis of the communication disaster, items (E) and (F) will be clarified. In addition, Sect. 2.5 will describe the international reaction to the Fukushima accidents.

2.2 Brief History of Nuclear Power Plants

How are nuclear power plants embedded in political, economic, and social contexts in Japan? Table 2.1 shows a brief time table of Japanese nuclear power plant development. Following the atomic bombs detonated in Hiroshima and Nagasaki in August 1945 and U.S. President Eisenhower's address on "Atoms for Peace" in December 1953, the budget plan for nuclear power in Japan passed the Japanese Diet in March 1954. In 1955, the Japanese Diet enacted the basic law for nuclear power, resulting in the establishment of the Science and Technology Agency in 1956.

In the 1950s, nuclear power was a kind of dreamy media for Japan's come-back story after World War II. As a result, in the Japanese political system, nuclear energy played an important role for post-war reconstruction and for overcoming Japan's limitation as a country of few natural resources. From 1956 to 1969, the government succeeded in siting nuclear power plants in 17 regions, and each of these power plants began operation at some point between 1970 and 2005 (Kainuma 2011, p. 298). In the construction process of nuclear power plants, the "dream for regional developments" by residents in the region and the "dream for independence of the resource

Table 2.1 Historical background of Japanese nuclear power plants (NPP)

Year	Event
1945	Atomic bomb in Hiroshima, Nagasaki
1953	"Atoms for Peace" U.S. Presidential address
1955	Basic law of nuclear power
1956	Establishment of Science and Technology Agency
1956–1969	Attempted siting of NPP by several municipalities
1970	Anti-nuclear movements in the world
1970–	Failure in attempted siting and in construction of NPP

supply of Japan" by the central government system led the two parties to cooperate. However, at the end of the 1960s and in the beginning of the 1970s, anti-nuclear movements prevailed all over the world and resulted in decreased support for nuclear power plants in Japan after 1970. This is a brief examination of the history of nuclear power plants from the period after World War II to 1970. In the next section, I will examine this period of history in greater detail based on the items I listed above.

2.2.1 Cultural Acceptance of Nuclear Energy

Japan was the first and the only country against which the atomic bomb was used to kill civilians. As a result, a "dark shadow" clouds the image of the "atom" for Japanese citizens. Cultural acceptance of nuclear energy in Japan is complex, and it can be divided into three phases; phase I (1945–1969), phase II (1970–2011), and phase III (2011–).

2.2.1.1 Phase I: 1945–1969

With the potential promotion of nuclear power in the 1950s, some physicists played an important role in garnering cultural acceptance of nuclear energy. For example, Koji Fushimi insisted on three principles, "peace," "openness," and "democratic control," for the promotion of atoms for peace, which became the Nuclear Power Charter by the Japan Science Council in 1952 (Yoshioka 1999). Another physicist, Taketani (1952), insisted that "Japan is the only county that ever experienced nuclear devastation; therefore, the Japanese deserve a strong statement on nuclear power. The Japanese have a greater right to do research on atoms for peace than other countries." Taketani's claims divided the use of nuclear power into two faces, light and shadow, and stated that the depth of the "shadow" from which the Japanese suffered from nuclear power gave the Japanese a right as well as a duty to use the "light" side of nuclear power (Yoshioka 1999). This claim to promote peaceful atom usage based on Japan's existence as a bomb victim strongly affected cultural acceptance of nuclear energy. The three principles mentioned above were included in the Basic Law of Nuclear Power (1955) with slight changes, appearing as "autonomy," "openness," and "democratic control." Raising these three principles, physicists persuaded the public of the need for nuclear power despite the public's anxiety regarding the negative side of nuclear power.

2.2.1.2 Phase II: 1970–2011

By 1970, the government's attempted siting of nuclear power plants succeeded in 17 regions (e.g., Fukushima, Fukui, Kashiwazaki-kariwa); however, in the beginning of the 1970s under the influence of global environmental movements, anti-nuclear activities gained momentum in Japan, and many residents began to resist plant construction. As a result, Japan experienced a segregation of promoters and opponents of nuclear power plants after 1970.

Juraku (2013) characterized this segregation through the analysis of the concentrated siting of nuclear power reactors at a single site. Pronuclear government, facing many anti-nuclear activities after 1970, promoted strategically concentrating nuclear power reactors at the sites where residents had already accepted nuclear power before 1970. At such sites, "fundamental problems and issues, which would have hampered progress, were ignored, down-played, neglected or shunted aside" (Juraku 2013, p. 52). Instead, pronuclear supporters focused on the local economic benefit and development by subsidies (e.g., Dengen San-pou Ko-fu-Kin Seido, which means a law on electricity to provide subsidies to local governments that support the generation of electricity). At these sites, any problems posed were seen as being manageable; therefore, residents came to believe that problematic safety factors would never become critical issues. On the contrary, citizens who lived in different areas did not see those problems as manageable and did not believe that safety factors would never become a critical issue. In this way, segregation on the basis of safety issues arose between residents in sites with nuclear power plants and those without them.

This segregation gives us some insight to answer several questions. The first question is: how are nuclear power plants embedded in political, economic, and social contexts in Japan? Pronuclear individuals who had institutional politics to enhance economic development with subsidies at sites with nuclear power plants used strategic agenda-setting to successfully promote the safety statements at these sites, segregating between pronuclear and anti-nuclear citizens. With the segregation, nuclear power plants are embedded in political, economic, and social contexts in Japan. This explanation leads to an answer to the second question: under what kinds of relationships between science, technology, and society are such accidents produced? In Japanese society, anti-nuclear activities existed, but their power did not reach to the sites with nuclear power plants. The strong segregation between pro- and anti-nuclear power activities developed in parallel with segregation in statements on the safety of nuclear power. The accident occurred within this situation.

In the introduction to this book, I raised three questions: (1) Why did the "precautionary principle," which existed in Japan in the 1970s to govern environmental issues, not work in the field of nuclear power plants? (2) From the observation of administrative lawsuits, several experts pointed to the lack of public engagement in the administrative process in the initial approval of the construction of the plants (Fujigaki 2009). These points, which came to light after the controversy in the administrative lawsuit, were not utilized for nuclear power safety discussion after the lawsuit. Why? (3) Finally, although the relationship between science, technology, and society has led to new technology in various fields (e.g., food sciences, including genetically modified organisms, and information technology), it has had little effect on the historically-rigidly constructed relationships among them in nuclear power energy. Why?

The above explanation on segregation gave us some insights into these questions. First, the "precautionary principle" had an effect on environmental problems of chemical contamination, as seen in Chap. 7 on the Itai-itai disease case. However, this principle in the environmental field could not reach the sites with nuclear power plants, since atomic power was promoted in the "atoms for peace" context and any problems posed were seen as being manageable at these sites. Second, public engagement and the construction of the public sphere were not enough in the field of nuclear power plants because of the segregation mentioned above. Why were the points that came to light after the controversy in the administrative lawsuit—that is, the lack of public engagement—not applied to nuclear power safety discussions or risk communication after the lawsuit? The reason is the existence of segregation. For example, the Japanese government (specifically the Prime Minister) approved the establishment of the Monju nuclear power plant in the Tsuruga District, Fukui Prefecture, in May 1983. In response, local residents began legal action against the government in September 1985. Although any problems posed were seen as being manageable at sites with nuclear power plants in areas that had accepted nuclear before the 1970s, the local residents who brought the Monju lawsuit did not believe the "manageable" or safety myth. Therefore, local residents who had legal action against the government in the 1980s did not have the same safety beliefs as local residents who accepted the nuclear power plants before 1970s. The former's skepticism did not reach to the latter's belief and could not deconstruct the belief. Third, the new relationship between science, technology, and society resulted in new technology fields like food science or information technology; however, the new relationship has had little effect on the historically-rigidly constructed relationship in the field of nuclear energy. The reason and basis for this "rigidness" is the segregation mentioned above. That kind of segregation makes mutual discussions impossible, and the "public sphere" for constructing new relationships can hardly become a reality.

2.2.1.3 Phase III: 2011–

The Fukushima Dai-ichi accidents have clearly revealed this kind of segregation. Now Japanese society is in the process of reconstructing the cultural acceptance of nuclear energy. At the same time, in a Fukushima health surveillance on the effect of radiation, doctors at Fukushima Medical University (FMU) are now facing many conflicts in risk communication which were partly caused by the segregation among the public. I will deal with this point again in Sect. 2.4.

2.2.2 Role of Nuclear Energy in the Political System

As I mentioned in the previous section, the budget plan for nuclear power in Japan, based on the U.S. presidential address on "Atoms for Peace" by Eisenhower on December 1953, passed the Japanese Diet in March 1954. In 1955, the Japanese

Diet enacted the Basic Law of Nuclear Power. Some politicians, including Mr. Nakasone who was the Prime Minister of Japan from 1982 to 1987, played an important role in promoting nuclear energy for peace and were also said to be considering atomic armament (Yoshioka 1999). The Nuclear Power Preparation Committee was established on May 11, 1954, and the Atomic Committee was established in January 1956. On May 19, 1956, the Science and Technology Agency (STA) was established and began conducting nuclear power research. At the same time, electric industries began to seek a way of constructing commercially viable nuclear power plants. From this point, the governance of nuclear power was conducted by the Ministry of International Trade and Industry (MITI) and the STA. Yoshioka (1999) insisted that the "dual-structured sub government system" in Japan began at this point. That is, Yoshioka (1999) asserted that policy decisions concerning nuclear power were monopolized by two insider groups: the alliance of the MITI and Japan's electric power industry on the one hand, and the STA on the other. These two insider groups in combination constituted a "sub government" outside of democratic control.

2.2.3 Status of the Nuclear Industry

The Japan Nuclear Power Industry Association was established in March 1956, and five groups of nuclear industries were established mainly by the heavy electric machinery manufacturers (e.g., Mitsubishi, Toshiba, Hitachi, etc.). The Kansai electric company established the Atomic Power Research Team (ART) in April 1956, and the Tokyo Electric Power Company (TEPCO) created the TEPCO Atomic Power Research Team (TAP) in January 1955. Genden (Japan Atomic Power Company) was established on November 1, 1957. In this way, in Japan, heavy electric machinery manufacturers and electric companies began to deal with nuclear power in the middle of the 1950s through their alliance with the MITI.

Following the beginning of the nuclear industry in the 1950s, the industry grew and developed in the 1960s. The Tokai Atomic Power Plants first went critical on March 4, 1965, and began commercial service on July 25, 1966. Many pressurized-water reactors (PWRs) and boiling-water reactors (BWR) began commercial service in the 1970s. For example, Tsuruga Daiichi opened in March 1970, Mihama Daiichi in November 1970, Fukushima Daiichi in March 1971, Mihama Daini in July 1972, and so on. In the 1970s, a total of 20 power plants began commercial service.

The First Oil Crisis in 1973 invited economic disorder in Japan because the Japanese economy depended on thermal power generation, which requires oil. In 1974, Dengen sanpo (the Law on Electric Power) was enacted to de-concentrate the risk of dependence on oil. The crisis and this law promoted the development of nuclear power. The law established certain amounts of subsidies to local governments at the sites of nuclear power.

The subsidies also included other power-generation methods, such as hydro-electric generation, wind-force power generation, and geothermal power generation. However, the amount of power generated through nuclear energy was so high that most of the subsidies were fulfilled by nuclear power. In addition, the monopoly of big electric companies also created obstacles for the promotion of other power generation sources. If electric power can be supplied by many small companies based on a change in law, then utilization of other power generation would increase. In this sense, several nuclear industries, especially electric companies, are given preferential treatment by the Japanese government.

2.3 Politics of "Beyond Assumption"

In this section, I will analyze the politics of risk-governance by examining the politics of "beyond assumption" discourse, which is often used in reports on accidents (Fujigaki and Tsukahara 2011). These analyses will help illuminate item (D), the source of legitimate expertise in this domain.

For examining this discourse, I will explain the detailed process of the actual disaster. An earthquake at 15:42 on March 11, 2011, triggered a large tsunami along the east coast of Japan, which damaged the cooling system of the Fukushima-Daiichi nuclear power plant and led to a hydro-explosion of the plant's core. After this accident, the words "unexpected" or "beyond assumption" were used frequently by nuclear engineering experts and the media. What is the meaning of "unexpected"? This word contains highly political nuance.

The Japan Nuclear Energy Safety Organization had already released a simulation report predicting the "loss of electric power supply of the cooling system" five months before the earthquake (Japan Nuclear Energy Safety Organization 2010). Table 2.2 indicates a comparison between the results of the simulation and what happened in reality based on Makino's analysis (2011).

Table 2.2 Comparison between results of simulation and what happened in reality

Results of simulation	What happened in reality
	March 11 14:46 Earthquake
Loss of power	15:30 Loss of power
	16:36 Damage in cooling system
	18:00 Fuel rod exposure
2.4 h later Nuclear fuel rod fall	19:00 Nuclear fuel rod melt
3.3 h later Damage in pressure container	19:50 Rod fall down
	March 12 00:49 Abnormal pressure of container
16 h later Breakage of pressure container	06:50 Melt down of fuel
	14:30 Vent from container
	15:36 Explosion

The simulation shows that, after a loss of power, the following would happen: at 2.4 h after the loss of power, the nuclear fuel rods would begin to fall. At 3.3 h, the pressure container would begin to show damage, and at 16 h, the pressure container would break. In reality, the earthquake occurred at 14:46. Power loss occurred at 15:30. At 19:00, 3.5 h after power loss, the nuclear fuel rod began to melt, and at 19:50, about 4 h after power loss, the rod began to fall down. At 6:50, almost all of the fuel melted. Thus, there is a clear correspondence between the simulation and reality.

Therefore, the "loss of electric power supply of the cooling system" was predicted, but no countermeasure was considered for this eventuality. In addition, the disaster produced much discourse asserting that the "loss of electric power supply of the cooling system was beyond assumption". Why were such statements frequently used when addressing the public? One reason is that professionals and the government were trying to shift the blame away from technology and TEPCO by saying that the situation was uncertain. However, the report on this disaster published in July 2012 by the National Diet indicated that the "loss of electric power supply of the cooling system" was predicted:

> Since 2006, the regulators and TEPCO were aware of the risk that a total outage of electricity at the Fukushima Daiichi Plant might occur if tsunami were to reach the level of the site. They were also aware of the risk of reactor core damage from the loss of seawater pumps in the case of tsunami larger than assumed in the Japan Society of Civil Engineers' estimation (National Diet Official Report of Fukushima Nuclear Accident Independent Investigation Commission, Executive Summary 2012, p. 16).

The reports by the Cabinet Office published in July 2012 also mentioned this assumption:

> The words "beyond assumptions," broadly speaking, can refer to two meanings. One means that an incident, which could not be predicted even with possession of the most advanced academic knowledge, occurred. The other one means that, in light of financial limitations and other limitations to the ability to respond to all predictable events, a line was drawn to exclude incidents that were realistically assessed to have a low probability of occurrence, and an incident of a scale far beyond that line occurred.
> Based on the study of the seismological progression and emergency preparedness administration over the past ten or so years, it is clear that the latter meaning held true in the case of the latest major tsunami (Cabinet Office Investigation Committee on the Accident at the Fukushima Nuclear Power Stations, Final Report, Executive Summary 2012, p. 30).

Thus, two reports indicated that "loss of electric power supply of the cooling system" was predicted and assumed. In particular, the second meaning presented in the Cabinet report (2012) is important: "in light of financial limitations and other limitations to the ability to respond to all predictable events, *a line was drawn* to exclude incidents that were realistically assessed to have a low probability of occurrence, and an incident of a scale far beyond that line occurred" [emphasis added].

We can re-consider "a line" in this sentence using the framework of risk-concept developed by Beck (1986). Beck published the book *Risk Society* just after the accident at Chernobyl. He divided scientific rationality from social rationality.

Scientific rationality deals with scientific probability and predicts hazard. On the other hand, to plan the countermeasure, we have to consider "what should be protected," e.g., the health of citizens, the environment, or the economic system with sustainable development. If we consider something that should be protected, then "a line" will be drawn, and based on this line, "probability" turns into "risk" concept. Based on this discussion, "loss of electric power supply of the cooling system" was predicted in terms of scientific rationality; however, it was unexpected and "beyond assumption" in terms of social rationality. Therefore, based on the expression in the Cabinet report, "beyond assumption" refers to an area of social rationality. "A line" was drawn "to exclude incidents that were realistically assessed to have a low probability of occurrence, and an incident of a scale far beyond that line occurred"; in other words, this line was drawn to protect mainly the economic system with sustainable development (see, e.g., "in light of financial limitations" in the Cabinet report) rather than the health of citizens or the environment.[1]

These facts indicate that in Japan prediction regarding scientific rationality is done adequately; however, integration of knowledge regarding social rationality has had some problems. In other words, the levels of research in nuclear technology and simulation technology as well as the levels of research regarding tsunamis and earthquakes in Japan were not low. However, this research and these technologies were not integrated for risk-prevention. In reality, there was no "sphere" to discuss this integration (Imada 2014). We have to admit that there is segregation not only between the sites, but also in fields of research. In addition, the lack of democratic control under the sub-government system, which I mentioned in Sect. 2.2.2, invites a situation in which engineers have been insulated from close investigation by the public. The American historian Porter (2013), who studied severe public scrutiny in flood control in the U.S., indicated that the "Japanese nuclear engineers were insulated to a striking degree from public scrutiny of the sort faced by American ones" (Porter, Preface for the Japanese-Translated Version of "Trust in Numbers" 2013).

We can consider this situation further by applying Bijker's (2007) comparative analysis of flooding in the U.S. and in the Netherlands. He analyzed the aftermath of the flooding of New Orleans by hurricanes Katrina and Rita in 2005, comparing Dutch coast engineering, and noted:

> Does this suggest that the US Army Corps of Engineers is less able than the Rijkswaterstaat engineers in the Netherlands? I will argue that something else is going on: that the

---

[1] In addition, there is a computer simulation that can calculate the development of an accident in real-time using the same code of the Japan Nuclear Energy Safety Organization (JNES). This also means that the accident was predicted in scientific rationality. Prediction sometimes becomes the cause of other victims. For example, as the Tsunami countermeasure, some professionals have done "disaster drills" in the Kamaishi-city based on their "assumption" of the effect of a tsunami. However, the height and power of the tsunami was beyond their assumption, and more than 50 people died even though they followed evacuation instruction by these professionals (NHK 2011, March 21). However, if we cannot make assumptions, then we cannot prepare for disasters. It was a criticism on what can we formulate the responsibility of these professionals.

difference is not one of expertise and competence. …I compare the styles of US and Dutch coastal engineering, and argue that they express different conceptions of risk management in relation to flooding. These differences can, perhaps, be explained by reference to the wider technological cultures of both countries rather than to the specific engineering culture (Bijker 2007).

Bijker finally indicated that the difference in technological cultures exists in the "risk criteria" and in the way to establish this criteria in the society:

> The risk criterion that is used in designing levees and other coastal defence structures in the USA is a 1:100 chance, or a "hundred year flood." This criterion is a technical norm, carrying important professional "weight" among coastal engineers, but it carries no legal authority. …in the Netherlands, … the water should be kept out. In the Deltaplan Law, the criterion of 1:10,000 was specified: not merely as a technical norm, but as an obligation embedded in the "Delta Law," unanimously approved by parliament (Bijker 2007).

We can apply Bijker's comparative analysis to the nuclear power plant accidents in Japan. The risk criterion of the tsunami that would lead to "loss of electric power supply of the cooling system" was not specified in the Japanese law approved by parliament. Rather, a closed community of engineers decided the criterion, and it was not exposed to public scrutiny, e.g. Diet deliberation. This practice of closed-community decisions is the technological culture of Japan. Mr. Kurokawa, who headed up the report of the National Diet, stated, "It is a man-made disaster," as an expression of this technological culture (National Diet Official Report of Fukushima Nuclear Accident Independent Investigation Commission 2012); however, this statement is fraught with controversy on "techno-orientalism" among researchers in an international conference. I will discuss this point again in Sect. 2.5 again.

In this way, the "beyond assumption" discourse reveals embedded politics to shift the blame for technology and TEPCO to "the situation under uncertainty." Segregation between fields of research, lack of democratic control and of public scrutiny, and the technological culture in Japan were exposed to the light of day by the accident. What is brought by this exposure? The aftermath is the fallen credibility and the public's lack of trust in authorities (both government and professionals), which leads to a "communication disaster." I will deal with these results in the next section.

2.4 Effect of the Accident on the Technology-Society Relationship in Japan

In this section, I will deal with communication between experts and citizens after the disaster. This analysis will make clear (E) the role of the media and (F) the culture of public debate over complex techno-scientific issues.

2.4.1 Communication Disaster and Enhancement of Segregation

After 2011's Triple Disaster—that is, the earthquake, the tsunami, and the nuclear power plant accident—Japanese society experienced a "communication disaster." The National Diet Official Report of Fukushima Nuclear Accident Independent Investigation Commission (2012) also stated that there was a communication failure in protecting public health. It indicated that sufficient risk-communication on radiation was not provided to residents. The information disclosure after the accident was not a good example for other democratic countries. American anthropologist Hugh Gusterson criticized that the "Japanese Government continues to announce disorganized knowledge.[2]" Chapter 3 of this book will deal with the detailed information on this communication disaster from the standpoint of the science media center.

A gap arose between the information that citizens wanted to know and the information professionals wanted to provide. Citizens who lived in Fukushima wanted to know impartial, non-partisan, broad information (Yamaki 2011); however, professionals wanted to provide decisive action guidelines and limited, absolute information. It is not a deception but a simple misunderstanding of what the public wanted. Professionals hold an ideal that what is most needed from the public to the professionals is to provide the public with decisive action guidelines and limited, absolute information. The Science Council of Japan insisted on "unique" or "unified" knowledge (Onishi 2012). The Japanese government and professionals were hung up on unique, decisive action guidelines and disclosed only "safety" information. As a result, Japanese citizens began to distrust the government and professionals. Likewise, the two groups experienced differences in the anxiety they felt over this information. Citizens (or residents) experienced anxiety over both the limited information and their distrust. However, professionals had anxiety over releasing information in a non-unified voice and for the public unrest.

These gaps raise important questions on the responsibility of scientists. Which behavior is responsible on the part of scientists: to disclose only unique knowledge decisive enough for action guidelines or to disclose a variety of knowledge and enhance the individual decision-making ability of the citizens? This is a very difficult question that can also be applied to item (E), the role of the media.[3]

To examine these questions on responsibility, we should think about the segregation in Japanese society. First, I will explain the salient value similarity

---

[2] It was a criticism in a joint plenary of the History of Science Society (HSS), the Society for History of Technology (SHOT), and 4S on Fukushima in November 2011 in Cleveland.

[3] Of course, we should distinguish between communication during "emergencies" and communication during normal life. However, the information disclosure attitude during an emergency is affected by the relationship between science and society as well as by the public's trust in authorities throughout the course of their normal life.

(SVS) model in social psychology. In social psychology, it is said that two components are necessary to construct "social trust." One is competence, and the other is fairness in motivation. Competence means ability, experience, and qualification. Fairness in motivation means impartiality, integrity, and honesty in motivation of research. If a person with competence does something with fair motivation, then people will trust him/her. This was a theory, but after the Triple Disaster, some statements made by people with competence and fair motivation were not trusted. Why? The SVS model will explain these situations better.

The SVS model postulates that shared values determine social trust in institutions and persons related to a technology (Siegrist et al. 2000). In this model, if an individual thinks that the person in front of him/her shares similar salient values with him/her, then he/she will trust that person. Therefore, one who holds the salient value to ease the public's worry trusts other people who hold the salient value to ease the public's worry. Likewise, one who holds the salient value to open neutral data trusts other people who hold the salient value to open neutral data. The same holds true for those who wish to abolish nuclear power. In this way, one trusts people who hold similar salient values, and this tendency accelerates the segregation of groups that have different salient values.

As previously discussed in Sect. 2.2, segregation between sites with nuclear power plants and other sites without them already existed in Japan. In the sites where residents had already accepted nuclear power plants before the 1970s, any problems posed were seen as being manageable; therefore, residents believed that problematic safety factors would never become critical issues, since they were manageable.

How did the segregation between sites develop or change after the accidents? First, new segregation developed within the sites with nuclear power plants. It is not hard to imagine how these residents felt when their belief in safety was shattered after the nuclear power plant accidents. Some people lost trust in authorities (engineers and policy-makers), while other people tried to keep their trust in authorities. One who holds the salient value to want to know impartial, non-partisan information trusts other people who hold the salient value to want to know impartial, non-partisan information. On the contrary, one who holds the salient value to ease the public's worry trusts other people who hold the same salient value. This kind of segregation was pushed forward by their decision-making on whether to stay in the land of their birth or to evacuate to other prefectures.

At the same time, residents who lived in sites without nuclear power plants also lost trust in specialists and the government since these authorities released only one-sided safety information after the accidents. One who holds the salient value to abolish nuclear power trusts other people who hold the salient value to abolish nuclear power. This distinction accelerated the segregation between sites with nuclear power plants and sites without them. Social trust toward international agencies [e.g., International Atomic Energy Agency (IAEA)] was also divided: some people clung to the hope that the IAEA would bring the Japanese government toward the right direction. On the contrary, others criticized the IAEA as the organization that enhanced the nuclear energy generation (Shimazono 2013; Watanuki 2012). In this way, the communication disaster after the accidents accelerated the segmentation of Japanese society.

This fragmentation of the society makes attempts to survey Fukushima residents' health difficult. In doctor–patient communication, both parties can easily share the salient value to fight the disease or to improve the quality of life. However, in doctor–public communication, there are so many different salient values, and doctors seldom share their salient values with the public. For example, in a crisis, it is the doctors' tendency to try to avert panic or to ease the public's worry, while the public—presumably suffering from radiation—wants to abolish nuclear power. Thus, in doctor–public communication, the doctor and the public rarely share salient values; therefore, it is very difficult to build trust among them. Several doctors in the FMU noted this situation (FMU-IAEA International Academic Conference 2013, 2014).

2.4.2 Culture of Public Debate Over Complex Techno-Scientific Issues

Under this segregation, Japanese society tried to engage in public debate over the decision-making for future energy sources, giving rise to a few questions, such as: How should Japan manage the future energy supply? Should we live with less electricity without nuclear power plants? What are some alternative energy sources?

The government of Japan (Democratic Party) tried to deal with questions of the future of energy using a deliberative poll (DP) survey. On June 30, 2012, the Agency of Energy Rescurce in the Ministry of Economy, Trade and Industry released their plan. The agency insisted that there is a strong need for nationwide discussion by citizens. So, in July and August 2012, the agency conducted the DP regarding future energy. From July 7th to the 22nd, they conducted random sampling and selected 6,849 people for the survey. Among these 6,849 people, 285 people were selected as candidates for the deliberative meeting. The idea of the DP is to conduct a poll before and after the meeting to see what changes were made in opinions.

Participants were asked to choose which of the following situations they would most like to see: 0 % nuclear power in the future, 15 % nuclear power in the future, and 20–25 % nuclear power in the future. On August 4th, the results of the pre-meeting polling were released: 42 % selected the scenario of 0 % nuclear power in the future. In addition, on August 5th, the results of the after-meeting polling showed that 47 % of people selected the scenario of 0 % nuclear power in the future. This result was released in the press on August 22nd, and the agency concluded that "Japan should set a goal of 0 % nuclear power by the 2030s." However, the Cabinet decision did not include this proposal in the body, but included it as a reference The precise results are shown in Chap. 5 of this book.

The Japanese DP attempt was widely broadcasted outside Japan. A researcher at the International Joint Conference of European Association of Science and Technology Studies and Society for Social Studies of Science (2012) remarked that the "earthquake and Fukushima accidents are changing Japanese public policy," and another researcher at the same conference stated that "Fukushima's case has become the trigger in Japan to re-write boundaries between 'public' and 'administration'." How can we describe these situations? Is it accurate to say that Japanese society is moving from "paternalism" to a "democratic society with public engagement"?

Paternalism in Japanese society means that people rely extensively on professionals and only want to know decisive action guidelines and limited, absolute information. Extensive public trust in professionals co-exists with the professionals' desire to provide unique information. At the same time, it co-exists with a lack of democratic control over the nuclear power "sub-government" model mentioned in Sect. 2.2. Furthermore, it invites a situation in which engineers are insulated to a striking degree from public scrutiny of the sort faced in other countries, as I mentioned in Sect. 2.3. After the accidents, the credibility of authority fell and trust in professionals was lost. In the aftermath, how should Japan construct new relationships between science, technology, and society, or between the public and professionals? Is there a right way to re-construct the trust exactly as it was before? Can Japan build trust which co-exists with the professionals' desire to provide decisive action guidelines and limited, absolute information? Can the country build trust which co-exists with a lack of democratic control and without public scrutiny? The answer may be no. Another way for Japanese society is to construct new relationships between citizens and professionals without extensive reliance on professionals, with democratic control, and with public scrutiny.

Some Japanese professionals have noticed and are examining these issues. Some committees of the Science Council of Japan are trying to summarize reports on free, unconstrained information disclosure by professionals. One of the committees self-criticizes the professional communities' silence, isolation, self-regulation, and information control in emergencies after the Triple Disaster, noting that they did not give enough information to the public (Science Council of Japan, Committee on Autonomous Disclosure of Wisdom from Scientific Community for Crisis Response, 2014 June 4). It is trying to enhance information disclosure without strict control and to provide plural opinions. This change means that professionals should not stick to a "unique" voice, but should provide impartial, non-partisan (independent from the government), broad information that the public can evaluate for themselves.

These discussions lead to conflicts between areas that scientists should decide and areas that should be open to the society. People who believe that professionals should disclose only unique knowledge that is decisive enough for action guidelines have insisted that professionals should decide what should be done. However, people who believe that professionals should disclose a variety of knowledge and enhance the individual decision-making by citizens insist that the public should decide what should be done. In other words, these people claim areas that we should open to the public. The former attitude is a classical, formal belief that professionals should disguise and hide all controversy from the public. M. Rudwik explained this attitude as follows:

> The role of formal published papers in relation to informal argument during the controversy could aptly be compared with the role of occasional—and generally unrevealing—press releases during the real hard work of diplomatic negotiations behind closed doors (Porter 1995, p. 220).

This metaphor is very insightful in the current situation if we change "published papers" to "information disclosure by professionals."

Information disclosure under disaster can expand to the discussion on information disclosure under uncertainty. What kind of knowledge should be disclosed under uncertainty? In the case of the earthquake in L'Aquila in Italy, scientists who disclosed only safety information were arrested and received prison sentences in October 2012. Therefore, knowing where to draw the line between disclosing only unique knowledge decisive enough for action guidelines or disclosing a variety of knowledge and enhancing the individual decision-making by citizens is very difficult. The legal responsibility of information disclosure (or no information disclosure) is in the process of discussion in the above committee of the Science Council of Japan. It is asking the system not to hold researchers personally responsible.

The DP attempt also contains a discussion on information disclosure under uncertainty. The decision to conduct the DP reflects the change in Japanese society from being closed and paternalistic (with high reliance on professionals without democratic control and without public scrutiny) to democratic (with public engagement in decision-making and with public scrutiny). However, as mentioned above, extensive segregation exists between sites as well as research fields; therefore, some people like paternalism while others like the democratic process. The culture of public debate over complex techno-scientific issues in Japan requires some sort of integration of this segregation.

2.5 Universal Lesson from Fukushima or Techno-Orientalism?

The Fukushima nuclear power plant accidents and the Japanese DP attempt have invoked many international reactions. For example, a Dutch sociologist at the International Joint Conference of European Association of Science and Technology Studies and Society for Social Studies of Science (2012) expressed that "people in European countries are curious about the future of Japanese nuclear power as well as about the effect of citizen movements on future policy." A French social economist at the same conference showed strong interest in the effect of Japanese nuclear power policy on similar policies in Europe. In addition, a German researcher at the same conference told me that it was an epoch-making event that the agency of Energy Resource, Ministry of Economy, Trade and Industry concluded that "Japan should set a goal for 0 % NP by the 2030s." Thus, Japanese policy and citizen movements seem to be attracting worldwide attention.

The Japanese accidents have induced two kinds of reactions in the world. First, based on the fact that the accidents happened in a high-technology society like Japan, the public has recognized the possibility that the same-level accidents could happen in every country. From this perspective, Fukushima is a "universal lesson" for every country, and human beings should question the safety of nuclear power plants. Based on this perspective, the German parliament decided to completely abolish nuclear power by 2022. For example, the German report by the Ethical Committee on Safety Energy Supply noted, "We consider it very important that the accident happened in a country with high technology like Japan" (Kumagai 2012). The report also explained that the German society cannot assure that a grand-scale accident will never happen in Germany. In this way, Japan was an important benchmark to judge nuclear power safety.

The other perspective is to perceive the accidents as a "made in Japan" disaster and to treat the accidents as "specific to Japan," rather than being universal. This perspective is supported by the statement of the chairman of the report of the National Diet, Mr. Kurokawa. In the "Message from the Chairman," he wrote, "What must be admitted—very painfully—is that this was a disaster 'Made in Japan'" (National Diet Official Report of the Fukushima Nuclear Accident Independent Investigation Commission, Executive Summary 2012, p. 9, line 15). This perspective induces the "techno-orientalism" which shifts the responsibility from universal technology to the Japanese techno-culture. If one adopts this perspective, then the accident is not seen as a universal lesson for every country with high technology country like Japan, but as a problem specific to Japan.

"Orientalism" is a term to show the biased image by Westerners of Eastern countries. This concept reveals that "general knowledge," which is considered to be non-political, has a tendency to have highly politically organized issues especially in literature, historiography, philology, and sociography. Saīd (1977) pointed out that, when knowledge is produced, there are several political conditions highly structured in circumstances of knowledge generation even if we cannot see them as follows:

> What I am interested in doing now is suggesting how the general liberal consensus that "true" knowledge is fundamentally non-political (and conversely, that overtly political knowledge is not "true" knowledge) obscures the highly if obscurely organized political circumstances obtaining when knowledge is produced (Orientalism, Introduction III 1977, p. 26).

In these political conditions, Western countries hold a biased view of the Orient (Eastern) countries. This biased view is called "orientalism.[4]"

---

[4] Orientalism is a term used by art historians and literary and cultural studies scholars for the imitation or depiction of aspects of Middle Eastern and East Asian cultures (Eastern cultures) by writers, designers, and artists from the West (Wikipedia 2014).

Science and technology are considered to be universal; therefore, they are left out of discussions of orientalism in literature, historiography, philology, and sociography. However, when we consider Kurokawa's statement on the "made in Japan" disaster, we must question the line drawn between the universal facets of science and technology and their culturally-dependent facets. In addition, Bijker's statement on "technological cultures" that I mentioned in 2.3 gives rise to the same question.

Scientific knowledge is considered universal, rather than culture-specific, and any scientists can produce scientific knowledge universally without regard to their nationality. For example, we do not distinguish between an electron in the U.S. and that in Russia or that in Japan. However, scientific activities to produce scientific knowledge are conducted by human beings whose institutions to support these activities vary in different countries. Likewise, the research environment, relevant laws, and historical backgrounds are vastly different among countries. It is very difficult to determine whether knowledge on risk management of science and technology is universal or culturally dependent.

One point of view is to classify the stage of science and technology: whereas science is universal, technology includes a culture-dependent facet. Furthermore, risk management includes deeper cultural dependency. This idea is well-used; however, placing too much emphasis on the cultural dependency of risk management causes us to blame the technological culture, not technology itself, when an accident occurs. If one considers that a grand-scale accident like Fukushima could happen in any other highly technological country, such as the U.S., France, or Germany, then the lessons from Fukushima will be important ones that should be shared with other countries. On the contrary, if one considers that the Fukushima accident was "made in Japan" and very specific to Japanese culture, then one can say that there is nothing that we can learn from Fukushima. Therefore, the decision of whether to consider the accidents as universal lessons from Fukushima or to regard the accidents as culturally specific on the basis of techno-orientalism determines the countermeasures in the policies of the next generation of nuclear power plants.

2.6 Conclusion

From the analysis of the processes through which nuclear power plants are embedded in political, economic, and social contexts in Japan, we can determine that segregation was established between sites that accepted nuclear power plants before the 1970s and sites without nuclear power plants. After the accidents of March 11, 2011, this segregation expanded between these sites as well as within each site. For example, those who hold the salient value to trust officials continued to support nuclear power while those who developed the salient value to distrust sought to abolish nuclear power. This segregation among the public with several different salient values and the historical segregation between the sites make Fukushima Medical University's health surveys very difficult.

In addition, discussions about whether to consider the accidents as universal lessons from Fukushima or to regard the accidents as culturally specific leads us to a discussion on technological culture with relevance to techno-orientalism. The remaining problems are: (1) how to build trust between doctors and the public in the health survey in Fukushima in light of the historical segregation worsened by their separation of salient values; (2) how to rebuild trust in professionals without extensive reliance on professionals and with public engagement in Japanese society; and (3) how to construct new relationships between science, technology, and society with democratic control and with public scrutiny in Japan. These are also challenges to the critical engagement by Science and Technology Studies (STS) with society.

References

Beck, U. (1986). *Risikogesellschaft auf dem Weg in eine andere Moderne*. Frankfurt am Main: Suhrkamp.

Bijker, W. E. (2007). American and Dutch coastal engineering: differences in risk conception and differences in technological culture. *Social Studies of Science, 37*(1), 143–151.

Cabinet Office Investigation Committee on the Accident at the Fukushima Nuclear Power Stations (2012). Final report. http://www.cas.go.jp/jp/seisaku/icanps/eng/final-report.html. Retrieved August 5, 2014.

Felt, U. (2013a). Presentation at the IAEA technical meeting, Vienna, May 6–10.

Felt, U. (2013b). Beyond refection: STS knowledge and practical action. In: Paper Presented at the FMU-IAEA International Academic Conference, Fukushima, November 21–24.

FMU-IAEA. (2013). International Academic Conference: Radiation, Health, and Society: Post-Fukushima Implications for Health Professional Education, Fukushima, November 21–14.

FMU-IAEA. (2014). International Academic Conference: Radiation, Health, and Population: The Multiple Dimensions of Post-Fukushima Disaster Recovery Fukushima, July 25–27.

Fujigaki, Y. (2009). STS in Japan and East Asia: Governance of science and technology and public engagement. *East Asian Science, Technology and Society: An International Journal, 3*, 511–518.

Fujigaki, Y., & Tsukahara, T. (2011). STS implication of Japan's 3/11 crisis. *East Asian Science, Technology and Society, 5*(3), 381–394.

Imada, T. (2014). Statement at the committee meeting. Kagakushya kara no jiritsutekina kagaku jyouhou no hasshin no arikata kenntou iinkai (Committee on autonomous disclosure of wisdom from scientific community for crisis response, Science Council of Japan). Summary of meeting minutes will be posted at http://www.scj.go.jp/ja/member/iinkai/jiritsuhasshin/jiritsuhasshin.html.

Independent investigation commission on the Fukushima Daiichi nuclear accident. (2012). *Fukushima genpatsu jiko dokuritsu kenshyou iinkai chyousa kensyou houkokusyo* (Independent investigation commission on the Fukushima Daiichi nuclear accident, final report). Tokyo: Discover 21.

Japan Nuclear Energy Safety Organization. (2010). Jishin ji reberu 2PSA no kaiseki: BWR (Level 2 PSA analysis for seismic events: BWR). http://www.nsr.go.jp/archive/jnes/content/000017303.pdf. Retrieved August 5, 2014.

Juraku, K. (2013). Social structure and nuclear power siting problems revealed. In R. Hindmarsh (Ed.), *Nuclear disaster at Fukushima Daiichi: Social, political and environmental issues* (pp. 41–56). New York: Routledge.

Kainuma, K. (2011). *Fukushima'ron: Genshiryoku-mura wa naze umaretanoka (Why was the "genshiryoku-mura" (nuclear power village) born?)*. Tokyo: Seidoshya.

NHK. (2011). *Close-up Gendai*. Broadcasted program. NHK, March 21.

Kumagai, T. (2012). *Naze Merkel wa tennkou shitanoka (Way Merkel changed her decisions?)*. Tokyo: Nikkei BP.

Makino, J. (2011). 97. Fukushima genpatsu no jiko (2011/3/19–21) (Accident at the Fukushima nuclear power plant on March 19–21, 2011). http://jun-makino.sakura.ne.jp/articles/future_sc/note098.html. Retrieved August 5, 2014.

National Diet. (2012). *Kokkai jiko cyou houkokushyo: Tokyo denryoku Fukushima genshiryoku hatsudenshyo jiko chyousa iinnkai* (National Diet official report of Fukushima nuclear accident independent investigation commission). Tokyo: Tokuma shyoten.

Onishi, T. (2012). Posuto 3.11 kawaru kagaku gijyutu rikkoku (ge): kagakusya no yakuwari wa; nihon gakujyutsu kagigi Onishi Tkashi kaichyou ni kiku (The role of scientists post-3.11: interview with Takashi Onishi, President of Science Council of Japan). *Nikkei Shinbun*, morning edition, p. 11.

Porter, T. M. (1995). *Trust in numbers: The pursuit of objectivity in science and public life*. Princeton: Princeton University Press.

Porter, T. M. (2013). Nihongo ban (2013 nen) eno jyo (Preface for the Japanese edition of *Trust in numbers: The pursuit of objectivity in science and public life*). T. M. Porter, *Su-chi no Shinraisei* (Y. Fujigaki, Trans.). Tokyo: Misuzu syobo.

Saïd, E. (1977). *Orientalism*. London: Penguin.

Science Council of Japan. (2014). Remarks by committee members at a committee meeting. Kagakushya kara no jirisutekina kagaku jyouhou no hasshin no arikata kenntou iinkai (Committee on autonomous disclosure of wisdom from scientific community for crisis response). Summary of meeting minutes will be posted at http://www.scj.go.jp/ja/member/iinkai/jiritsuhasshin/jiritsuhassh n.html.

Shimazono, S. (2013). *Tsukurareta housyasen 'anzenron' (Man-made radiation safety)*. Tokyo: Kawaide shobo shinshya.

Siegrist, M., Cvetkovich, G., & Roth, C. (2000). Salient value similarity, social trust, and risk/benefit perception. *Risk Analysis, 20*(3), 353–362.

Taketani, M. (1952). Nihon no genshiryoku kenkyu no houkou (Direction of the Japanese nuclear research). *Kaizou, 33*(17), 70–72.

Watanuki, R. (2012). *Housyanou osen ga mirai sedai ni oyobosumono (Chernobyl research: effect of radiation contamination on next generation)*. Tokyo: Shinhyouronsya.

Yamaki, T. (2011). Statement in the annual meeting. In: Paper Presented at the 10th Annual Meeting of Japanese Society of Science and Technology Studies, Kyoto.

Yoshioka, H. (1999). *Genshiryoku no shyakaishi (Social history of atomic energy)*. Tokyo: Asahi Shimbunshya.

Chapter 3
Agenda Building Intervention of Socio-Scientific Issues: A Science Media Centre of Japan Perspective

Mikihito Tanaka

Abstract In the present day, social agenda is constructed on media, but its manner has become complicated after the rise of the Internet. Socio-scientific issues also form their shape on media, so it could be said media is an origin where misunderstanding between experts and citizens begins. The theme of this chapter is to depict the difficulty of performing deliberation within this complex media ecosystem with subsuming socio-scientific issues. Furthermore, by figuring out its difficulty, this chapter will throw questions at Science and Technology Studies (STS) about its significance and role in the face of the media argument. In order to outline this complicated problem, this chapter takes the following configuration: First, depict the structural features of science media in Japan, which is an object of intervention of the Science Media Centre of Japan (SMCJ). Then we will describe the functional procedures of the SMCJ. This will lead to a reflexive examination of the SMCJ's activity following the Great East Japan Earthquake, where we will discuss the significance and limitation on intervening agenda building and framing processes in the media

3.1 Introduction

Even now after 3 years have passed since the Great East Japan Earthquake, Japanese society reflections reported by the media have been divided by the debate regarding risk. Among them includes a prominent gourmet manga series, where characters warn that "people in Fukushima get nosebleeds because of radioactivity from the nuclear accident", which lead science defenders to sneer at its irrational insistence (cf. McCurry 2014). Another example is the slightly increasing number of patients identified with pediatric thyroid cancer who find themselves trapped in an abstract political debate in the name of science; are the figures just in the error range, or do

M. Tanaka (✉)
Waseda University, Tokyo, Japan
e-mail: mikihito.tanaka@gmail.com

© Springer International Publishing Switzerland 2015
Y. Fujigaki (ed.), *Lessons From Fukushima*,
DOI 10.1007/978-3-319-15353-7_3

they have significant value, and represent the effects of tragedy? But, like always, the actual victims are shifted into the background of these agendas.

It looks like the social atmosphere has already been determined. Parallaxes between polarized clusters—the "scientifically judging" majority and the warning minority that have been visualized significantly by new forces such as social media—seem completely incommensurable; they are attracted to one another by their own ideologies, overlap in complex manner inside the media sphere, but are isolated, and incubate antipathies toward their opponents from within their own beehives. Experts and citizens, media elites and active audiences, central and local; there is a deep distrust among different stakeholders and the "scientific facts" locates between them. The Nation of Japan, which was built by its faith in this science and technology, is entering an era of confidence crisis that has never been encountered before.

This dis-communication depends a lot on the malfunction of social and media systems that existed before the disaster. But what on earth does a scientific debate regarding risk in the media after a disaster look like? The textbook defines it as "care communication", an intellectual triage that has been trained as knowledge by discussion based on the norms is one of the measures for such situations (Lundgren and McMakin 2011). However, such advance preparations do not always fit perfectly to each individual situation. It always produces invisible victims who have spilled out from the prepared net. Therefore, to help potential victims that have been marginalized, society must recognize them as stakeholders, and must keep repeating ad hoc "consensus communication" for reviewing the risk frame.

Whether we like it or not, this reframing is carried out within the media. If it comes to the distribution of socio-scientific risk, crucial stakeholders in this process are journalists and scientific experts. In addition in modern society, citizen "prod-users (Bruns 2008)" participate in this reframing struggle by using accumulating powers such as "click", "like" or "tweet".

During debates regarding science and technology deployed in the media, what can the academic discipline of science, STS studies do? Or, even if something can be done, will it cause any new problems? Through the operation of Science Media Centre of Japan, the authors tried to intervene the social discussion of socio-scientific debates in society following the Great East Japan Earthquake. By accident, we received high reputation within Japan. However, to question our own, we do not think that evaluation was appropriate.

The theme of this chapter is to depict the difficulty of performing deliberation within this complex media ecosystem with subsuming socio-scientific issues. Furthermore, by figuring out its difficulty, this chapter will throw questions at STS about its significance and role in the face of the media argument. In order to outline this complicated problem, this chapter takes the following configuration: First, to depict the structural features of the science media in Japan, which is an object of intervention of ours in the form of the SMCJ. Then, we will describe the functional procedures of the SMCJ. This will lead to a reflexive examination of the SMCJ's activity following the Great East Japan Earthquake, where we will discuss the significance and limitation on intervening agenda building and framing processes in media.

3.2 Background: Crisis of Credibility After 3/11

As a consequence, the Great East Japan Earthquake was a disaster-derived "crisis of confidence" for both science experts and journalists. According to a poll, the public were conscious of the importance of the media after 3.11, but its credibility dropped throughout the reports after the disaster (Hakuhodo DY Media Partners 2011). Evaluation of the Japanese media also fell internationally. Until 2010, Japan was ranked around 10th in the Reports Without Borders "Press Freedom Index", but rapidly moved down in ranking year by year after 2011. This position has been described as a "plummet caused by censorship of nuclear industry coverage and the country's failure to reform the 'kisha club' system (RWB 2014)". We will consider later the validity of this evaluation, but it is clear that Japanese media system was in malfunction after the challenge of the disaster.

On the other hand, it is said that credibility of the scientists among public also declined after the disaster. From the results of various polls, the Japanese Ministry reported in its white paper that "it seems that experts in general are not fully aware that the Japanese public's confidence in scientists/engineers has dropped, and that there is a sharp increase in the number of Japanese who believe that experts alone should not determine the direction of research and development in science and technology (MEXT 2012: 54)". These "not fully aware" experts are doubtful about the concept of "crisis of credibility" itself, and "the problem of lose credibility of government and related experts were hidden, and situation is depicted that essential problem is citizens' lost of reliance (Kageura 2013: 71)". Nevertheless, regardless of who is responsible, they agree that function of the media as an open forum malfunctioned after the disaster. This malfunctioned structure is none other than the subjects that we attempted to intervene artificially in the name of the SMCJ.

We should indicate that science credibility was certainly high enough in the pre-disaster Japanese society. That is, the "public understanding of science" function of media was working well in Japanese society. Because of this, there was excessive trust toward the system that was lost after the disaster.

In general, science experts in Japan do not think highly of their science media. Many of them highly regard foreign articles from outlets such as *The New York Times* or *The Guardian*. They sometimes complain that compared to foreign media, science journalism in Japan is lowbrow (NAOJ and SOKENDAI 2005; Segawa 2010). However, if you want to look into the details of studies about media, or in-depth interviews with those experts, it is difficult to agree with such a statement. Many of the issues noted by the experts are only the problems inevitably occurring when performing the irreversible translation from complex scientific knowledge into layman's terms. Most of problems stated by experts are in the range of inaccurate paraphrasing of science knowledge. Those statements miss the fact that Japanese science media is defining its audience as a whole nation, including the elite and ordinary people.

Thinking from the point of view of the diffusion and popularization of scientific knowledge, and especially generalization of a standpoint of the majority of the science community, Japanese media appropriately—or more than necessarily—reflects the balance of the majority opinion in the scientific community. For example, compared to the Anglophonic media, comments from skepticists rarely appear on global warming issues in quality press and public broadcast (Asayama and Ishii 2012; Nagai 2015); Complementary medicines are rarely treated in media contents compared to Canadian and Australian media, and basic framing towards them are cautious, and mentioned with strong contingent condition of their usage (Tsuchiya 2013). In the field of regenerative medicine, articles are reported from an internal viewpoint of science community, and socio-scientific problems such as ethics are just added to the end of the article in a sentence as a strategic ritual (Shineha et al. 2014). Compared to Anglophonic media, deviant or pseudo-science are rarely mentioned in national quality press or public broadcasts that have principal agenda setting functions.

Moreover, science in action, or political arguments might be newsworthy issues from the point of view of the media, especially considering the mission of journalism, but they are not mentioned in Japanese contents. For example, neither 'Climate Gate' nor arguments about genetically modified foods derived by Séralini's paper (Séralini et al. 2012) have appeared in Japanese media.[1] Of course, grasping the actual condition of *ignorance*—the massive power usage of journalism—is not easy. A fragment of the Japanese media custom appeared in the following testimony by a science journalist in national press: "We cannot make contents about the controversial issues in the world of science in the Japanese media; unless an actual Japanese victim is confirmed, or included in the political and economic issues, even if those topics are popular overseas (Segawa 2010; Supplement 35)". To be clear, this can be a variation of the fundamental problem in journalism: structuring news discourse inevitably creates polarization between *us* and *them*. However, when considering the popularity of Nobel Prize topics in Japan, or daily disaster reports in news from abroad with "unworthy victims" for Japanese, this ignorance seems to fangle.

What kind of media structure brought about these tendencies? Focusing on the newspaper media, it can be summarized into three points: centralization of power in macro structure of the media, meso-level structure such as "*Kisha-club* (press club)" and huge science section in national newspapers, and news formats as a micro level custom. In the following section, we will try to analyze this structure using Hallin and Mancini's terms (2004).

First, the attentive media network in Japan—which can be represented by the newspaper circulation number per household, 2nd to Norway in the world—is a centralized network where the vertices meet at the capital city, Tokyo. Television

---

[1] For example, "Climate Gate" which occurred in November 2009 was covered only in a single brief article on December, in liberal national newspaper, Asahi-Shimbun. In latter 5 years, only 7 articles in Asahi-Shimbun covered the issue and all of them are short trivial article such as AAAS meeting report from USA or book review. Conservative Yomiuri-Shimbun only reported 1 article in 2012. Articles picking Séralini's case were only 2 in Asahi in introducing Anti-GMO movie, and none in Yomiuri.

networks have an overwhelming effect on framing of news, but newspapers still have main agenda setting functions, and print press and broadcasting networks are often cross-owned. From the scope of the distribution area of the newspaper, it is possible to distinguish the hierarchy: national, block, and local newspapers. National newspapers and news agencies lead the social agenda, especially about science topics in local newspapers and commercial broadcasts. The origin of this accumulating structure was the one newspaper per one prefecture policy put in place during totalitarianism before WWII, but it still affects the media structure today.

At the meso-level, this accumulating structure is strengthened by the notorious *Kisha-club* system, which numbers more than 800 and is located in almost all government and municipal offices, and many companies. However, antipathy from a freelance journalist, which is reflected in the RWB report described earlier, is not essential. The essence of the problem of the *Kisha-club*, which was declared following the disaster, is not its exclusive membership but that it is the easiest place for gathering information. This environment allows journalists to sit in an armchair and easily tame watchdogs to lapdogs. "Packed journalism" is not a concept Japanese journalists can avoid. It is a fundamental discipline for press, and therefore, they gather at a *Kisha-club* to provide homogenized frontpages similar to the other media to their customers. After the disaster, the 'spin' of these old customs lead journalists to weaken under an information overload. As a result, journalists were easily allowed 'above the line' spin (Gaver 2001). For example, just after the disaster, journalists suspected that the nuclear reactor was melting down. But when TEPCO stopped using the word 'meltdown' in press conferences, which also suggests China Syndrome, and substituted it for more 'accurate' words such as 'possibility of damage on fuel cladding or reactor core', it caused the word 'meltdown' to disappear from the headlines (Segawa 2011).

Furthermore the most important factor in a meso-level structures is that science journalists in national newspaper have strong agenda setting capacities about science. Each national newspaper has a large-scale 'science desk' made up of 50–70 science journalists. But these science journalists are just expected to translate science for broad audiences. In the newspaper structure, their first loyalty is obligated to science itself, not citizens. That is, asking questions about social issues surrounding science is not their main task; it is a task for other journalists who belong to the general news desk. As a result, after the disaster, risks were reported in *separated commentary-style*: in a scientific article, perspective was entrapped in the normative bias of science and explained, persuade and advocate for rational understanding of risk. In general news articles, anxiety of citizens toward risk was depicted in lay words, but sometimes unscientific words. In this way, science and anxiety over the risk was divided.[2]

The format of the article is one with micro level factors. Japanese press tend to have *internal pluralism*—covering different opinions within one medium—rather

---

[2] According to a study of Yomiuri-Shimbun and Asahi-Shimbun, based on the discourse and content analysis from 11 March 2011 to 11 March 2012 (Qi 2014).

than *external pluralism*—covering different opinions within similar media system. Also, the Japanese press chose to be *neutral arbiters* rather than be advocates, and chose *information-oriented reporting style* over *opinion-oriented style*. They often advocate the cause of neutral, fairness and nonpartisan ground, but treat those concepts as a substantial trophy. Audiences require this attitude, and sometimes blame media from the point of view of each of their own ideologies, especially netizens (net-citizens) who point their finger toward the media as biased if the press make an opinion-oriented article. When these symptoms overlapped with the hegemonic idea of 'objective' science perspective, the whole media structure surrounding controversial socio-scientific issue takes on a fragile aspect. For example, when an article or a program treated socio-scientific risk that has low probability but high consequence in a precautionary manner, the content would bring under dispute that "the contents might be unscientifically biased". In the past, the arenas of these controversies were remained in small communities, and experts were just politely contempt for the media. But nowadays, netizens including experts directly abuse journalists. As a result, the extent of political orientation within media content has became low, especially in the socio-scientific issues.

At a micro and practical level, the fact that the basic format of the article is a separated commentary style rather than a blending commentary style derives an important meaning to the legitimization of scientific risk. At the beginning of the crisis, the care communication stage after the disaster, science journalism tried to understand the risk with an emphasis on views of scientists. But after the first shock, the situation gradually moved into the recovery stage where consensus communication was required. Despite the changing situation, layman's terms from citizens in Fukushima describing actual anxiety that victims were facing were treated in separated articles (Qi 2014). In this style, anxiety was categorized as irrational feelings in contraposition with scientific rationality, legitimized and diluted by other factors. For example, anxiety of farmers who tried to make rice in contaminated areas of Fukushima was described as a struggle to make low-level contaminated rice under the accepted value, and the socio-scientific politics behind the value was abstracted. Or, the story was arranged into a more caricatural manner such as a heroic story of farmers who stood up against radioactive risk that threatened *our* rice—a national identity for the Japanese (Xiong 2014).

3.3 Agenda Setting of Science in Japanese Media

In brief, this system or architecture of Japanese media is a consequence of autonomous development of on-the-job training program as a journalist. As previously mentioned, Japanese media was organized under totalitarianism, and a new norm was given from the US following WWII. But this subsumed a nurturing system within the media, not within educational institutions. Hayashi has pointed out that "such bureaucratic/organizational dominance of intelligence has inertia that substitute the 'skill' to 'mission'. Therefore there is a tendency to deny the open and active forum, which incubates recursive expertise of departure, innovation, reflection or skepticism (Hayashi 2011: 193)". This is a statement for the whole structure of Japanese journalism, but especially in *science* journalism located between two similar but different disciplines, where this trend is amplifying.

These structure and customs of the media were beneficial for public understanding of science during peacetimes. Namely, politically judged scientific legitimacy is efficiently spread out among public via media cascade such as *Kisha-club*. But this is just a diffusion of scientific knowledge, or just transferring the majority opinion in scientific arguments. Operating principles of science that encapsulate uncertainty is not shared in society. Japanese media have carefully avoided controversial problems until it became serious. In other words, quality press functions in Japan have been a strict and convenient gatekeepers *for* science. This architecture had set scientists up in classical Marton CUDOS-like idle.

The SMCJ was a project to intervene with such media architecture. When we were preparing for the establishment of the SMCJ, people overseas who knew enough about the Japanese science-media relationships often asked us why we needed an SMC in Japan. It is a quite natural question from the point of view of transferring scientific perspective. We already have an efficient system. Accordingly, its hegemonic existence with infallibility of science in media forum is also our subject to intervene. Our ambition is not to set agenda within science, but build agendas by engaging with society.[3]

3.4 The Success and Challenges of the Science Media Centre of Japan

3.4.1 What are SMCs?

The 1985 "The Public Understanding of Science" report published in Britain is considered to mark the beginning of the country's Public Understanding of Science movement, but its perspective was criticized as deficit model. A new "Science and Society" report in 2000 in wake of the BSE scandal which threatened public trust in science and politics (Royal Society 1985; Wynne 1991; Office of Science and Technology and The Wellcome Trust 2000) pushed for activities that could mediate with the media, which lead to the establishment of the world's first Science Media Centre (SMC) in 2002. With the slogan, "where science meets the headlines", the SMC would act as a press office for the scientific community, providing accurate scientific evidence for the media when scientific issues became headline news, in order to improve the quality of scientific debates within the public sphere.

---

[3] This concept of 'agenda building' is based on Lang and Lang (1983) and Protess and McCombs (1991).

More than 10 years have passed since its birth, and the SMC movement has established itself as a valuable resource for both journalists and scientists worldwide, with an additional five independent SMCs being established across the world in countries including Australia and Canada. Preparations are also well underway for new SMCs in the US and China. Through its expansion, a natural advantage also became apparent. While national science bodies are bound by policies, the SMC collects scientific evidence, something that can be shared with other SMCs, and therefore is a powerful tool when global scientific issues become local news. For this purpose, the SMCs met for the first time in Doha, Qatar, in 2011 to establish a set of ground rules for global activities.

However, although scientific evidence itself does not have any borders, the issues each SMC chooses to address are determined by its country's culture and history. It means that although each SMC is fundamentally the same, the way in which it engages with its scientists and journalists varies. This gives way to different challenges. For example, in the past decade the UK SMC has succeeded in its goal to become a go-to resource for the UK media, but some critics have pointed out that its success means it can also manipulate the science agenda. More and more, the same quotes provided by the SMC can be found dotted around in different media outlets throughout the UK. This phenomenon is similar to the criticism Japan's exclusive *Kisha-Club* system has been getting for years. In the US, it could be referred to as oppressing freedom of the press (Macilwain 2012).

Still, the actions of a single organization such as the SMC, or the actions of a single journalist cannot be labeled as the only cause threatening the media's agenda building mechanism. It is far more complex. For the past 20 years, the media environment has changed to one that is making more journalists turn to churnalism (Davis 2008). This includes a growing dependence on information packages such as press releases, which in UK is reported to make up almost 80 % of all articles (Lewis et al. 2008), and is similar to what is being seen in the Japanese media (Segawa 2010; Ooi et al. 2014). Journalists today find themselves competing in an environment where people are exposed to information and public relations 24 hours a day, and where the political economy is encouraging science to become more visible. How journalists and organizations like the SMC are supposed to adapt to this landscape remains a constant challenge for both parties.

3.4.2 Science Alerts: The Fundamental Quality of the SMC

With this background in mind, let us look at the SMC movement in Japan. The 1999 conference on "Science for the 21st Century: A New Commitment" in Budapest marked a turning point for government officials and scientists, and it steered the way for science policy development based on public engagement. In 2005, the Special Coordination Funds for Promoting Science and Technology was set up to train science communicators. Three universities were allocated 500 million yen over a five-year period[4] including Waseda University. With this funding, the university established a training program for science journalists, recruited projects to research the challenges facing science and the media, and set up a Master's degree education program to train a future generation of journalists.

In 2009, the Japan Science and Technology Agency (JST) Research Institute of Science and Technology for Society (RISTEX) provided funding for the author to investigate the relationship between the Japanese science community and the media. By October 2010, the Science Media Centre of Japan (SMCJ) was established thanks to a three-year funding grant allocated by RISTEX to develop a new method of engagement between scientists and the media. In 2013, additional funding was granted to examine the SMCJ's norm required in its action. Currently, the SMCJ is receiving funds from the JST Center for Science Communication while continuing to search for a way to maintain its sustainability.

The fundamental activity of the SMCJ is no different from the original SMC in the UK. It has built a database of scientists willing to engage with the public, and a database of journalists, both working for media companies and freelancers who are all motivated to provide the public with healthy scientific evidence. It monitors and reacts to social agendas, collects expert opinions on issues, and forwards these comments to the media.[5] However, unlike the UK SMC where the goal is to provide science with a voice in the media, the Japanese media already have access to large number of scientific voices. Rather the problem in Japanese media is building agendas in a bottom up approach.

For this sake, the SMCJ definition of a journalist is not limited to working journalists, but also extends to freelance journalists, and prominent science bloggers. A registration process has been put in place, and potential journalists are asked to submit some of their past articles in order to filter out any persons wanting to exploit the SMC's information, some of which is provided to journalists under embargo, rather than use it to contribute to healthy public debates.

One way the SMCJ provides the media with information on what other stakeholders want to hear about is through Science Alerts (SA). Before going into detail about the SA process, it must be noted that this process is unique to the Japan's media environment, and that overseas SMCs may carry out SA-making differently. The SMCJ is also constantly updating its process so the details outlined here are true at the time of writing this article.

SAs largely consist of three steps. The first is to identify the topic. Monitor news agendas, discussions or social media networks, scientist comments, and SAs

---

[4] The three programs are as follows: The University of Tokyo established the "Science Interpreter Training Program", for sub major about communication and STS program for graduate students in natural science course. Waseda University made a graduate school for science journalism, the "Master of Arts program for Journalism Education in Science and Technology (Currently named the Graduate School of Journalism)". Third and h most successful was Hokkaido University's "Communication in Science and Technology Education and Research Program", which was open to every citizen.

[5] In other SMCs, the same service as Science Alert is called a Round Up or Rapid Reaction.

coming in from oversea SMCs. Topics are debated during team meetings, and outside scientists and sometimes STS scholars are consulted before a topic is selected. This is possibly the most important step in the SA process. It is the moment when the SMCJ has to decide whether a topic is potentially controversial, whether it poses a threat to someone such as potential victims after unfair distribution of risk, or if it a topic contains a hidden scientific message the mainstream media might miss.

What gives the SMCJ the right to decide what issues are most relevant to society is a constant issue the SMCJ faces, and the team constantly updates the way it works, and documents the steps it takes in order to maintain transparency.

Once the topic is decided, the next step is to identify potential experts to comment on the topic. The SMCJ database is used to list up potential experts capable of commenting on the topic, and research paper databases are used to identify any new experts. Media and social media comments are also searched to identify any other potential experts who might be willing to engage with the public. Experts are also valuable consultants who can provide local knowledge on the top scientists in their fields.

The final step is to contact the experts by email or phone to negotiate whether they can provide the media with a short comment of their point of view on a topic. Comments can range from a few lines to three paragraphs, which are put together by the SMCJ staff, many of who have science or journalistic backgrounds. Most comments begin with stating what the current situation is from the point of view of canonical science, then point out the issues with declaring the commentator's position in scientific argument, and finish with personal opinion and advocacy—what this all means to the public. At least three expert comments are collected in order to provide the media with a variety of views under performing triangulation of the scientific opinion, and the comments themselves are fact checked by the experts themselves before being sent to the media. In all cases, experts have provided comments for free as they see it as part of their responsibility to engage with the public, whose taxes are the main source of funding for the majority of scientists in Japan.

One problem the SMCJ does often encounter is when an expert goes beyond the line of providing scientific evidence, and begins pushing their own ideas or agendas into the argument.

The finished information packages are sent via email to journalists registered to the SMCJ, and are then uploaded to the SMCJ website after a few days.[6] Ongoing issues are constantly updated, and new comments are collected if required.

Following a survey of scientists, the SMCJ learned the biggest fear scientists in Japan had of the media was the risk of being mis-quoted.[7] The advantage of the website is that it acts as a safety net for scientists. If, for example, an expert is misquoted in a story, the expert is able to use the SMCJ website as evidence for his or her claim. In this sense, it fulfills a role of watching the watchdog.

---

[6] This is because important papers related to social agenda is sometimes restricted by embargo.

[7] In general, results were similar to other surveys such as in USA (Pinholster and O'Malley 2006).

While it is difficult to measure the impact the SMCJ has been making in the Japanese media, there are a number of case studies which have yielded visible effects. One of the first SAs produced by the SMC in November 2010 was an information package alerting the media on an upcoming collaboration between a major Japanese education company and a Chinese gene testing company. The collaboration was part of a new initiative to introduce gene testing to parents wanting to find out potential talent within the DNA of their children, and to use it to pinpoint what educational material would be most useful for their child. The SMCJ collected comments from five experts including a geneticist, molecular biologist, a bioethics researcher and a STS expert. The aim of the SA was to provide the media with scientific evidence and uncertainty from the five experts on the current stage and challenges of gene testing, and whether the technology available today is enough to determine the educational needs of a child. The SA was utilized by mainstream media including national daily newspapers Yomiuri Shimbun and Asahi Shimbun, and was retweeted more than 400 times. In fact, the effect was so much that the collaboration between the two companies in question was put to an end.[8] Cases such as this are a reminder that the SMCJ does have potential in Japan, and it can pick up on social issues traditional media outlets are not familiar with yet.

3.5 SMC Activities Following the Great East Japan Earthquake

At precisely 14:46 local time on March 11, 2011, the Great East Japan Earthquake struck the east coast of Japan. The disaster caused Fukushima Nuclear Powerplant's accident, which later became a typical example of "natural-hazard triggering technological disaster (Natech)". The disaster also became a complicated and meaningful media event with historical and cultural importance, which can be compared with Lisbon earthquake in 1755 (cf. Pantti et al. 2012).

While fallen books and papers scattered the SMCJ office floor, the team fixed up their computer screens and got to work. Even though the SMCJ had only been open for 4 months and we had little experience, the team managed to put together a Science Alert containing scientific information about earthquakes on the same day. We continued to monitor and act on the scientific debates that were unfolding within the nation, and collected expert comments almost around the clock. We did all we could to provide the public with what information they needed.

During this time, the SMCJ largely carried out the following (cf. Namba et al. 2014). First, we introduced journalists to experts and information sources. Second, we developed a FAQ page. Third, we put together and sent out SAs. Finally, we

---

[8] In this case, SMCJ might perform excess function than expected. We did not expect to stop the service. The SMCJ was misunderstood for a media company because of this first case. In addition, nowadays there are many similar services provided by education companies in Japan.

worked together with SMCs around the world to collect information. The details of each activity are described in the below sections.

3.5.1 Helping Journalists

Most of the enquiries sent to the SMCJ were from overseas journalists looking to interview Japanese scientists. They had heard about the SMCJ through other SMCs overseas. While we were able to introduce experts to these journalists, very few Japanese journalists approached the SMC, most likely because they had no knowledge about how to 'use' a SMC. Nonetheless, the SMC did receive a handful of enquiries from local journalists, including one from an economics magazine writer who contacted the SMC in search of a scientist who could explain what a Geiger counter was. The economics journalist had noticed that Geiger counter sales had been going up shortly after the nuclear power plant accident, and was looking for someone with a scientific background to explain the science to him. Around the same time, experts were cautioning on SNS about the problem that a number of individuals had been using Geiger counters incorrectly. Members of the public had been pointing Geiger counters directly at their food and judging for themselves about whether it was contaminated or not. While some vegetables such as spinach that contain potassium do give off a radiation reading, a scintillation detector is needed to make an accurate reading of other produce. In other words, Geiger counters are unreliable when used on most foods. By connecting this journalist with an expert who was able to explain these details, a story based on the needs of the public was published warning people about improper use of Geiger counters.

The SMC continued to provide the media with help on topics the public were demanding. Learning from cases such as Minamata disease or AIDS, which were identified by Japan's science history and STS groups, we had been able to offer the research results of experts through story-telling.

3.5.2 Putting Together a FAQ Section

Following the earthquake, the biggest impact the SMCJ made was creating a FAQ section regarding the nuclear power plant accident. Uploaded to the SMCJ's website on March 12, and constantly updated through to March 19, the information within the FAQ was based on people's voices observed on Twitter. By monitoring tweets containing keywords such as "radiation", "atomic", "tsunami", and "exposure", and following specific hashtags, the team was able to follow what was happening inside the disaster zone.

Twitter also enabled the SMCJ to identify scientists trying to provide the public with answers. Ryugo Hayano, a physics professor at the University of Tokyo was one expert the team noticed was gaining popularity as a hub for information. The SMCJ selected a number of those discussions and issues that had been unfolding and collected them into one FAQ page. The FAQ collected information on radiation exposure, about cooling processes in nuclear reactors, boric acid, hydrogen explosions, and many other components in a nuclear reactor. The information in the FAQ was constantly updated as more and more researchers on social networks started to voice their own opinions. For almost a week, more answers were added to the section, some with different views from others. Given that our goal was to provide perspective from a diverse group of experts, the team labeled these answers as A1, A2, etc.

We tried to make science advocacy clearer through the FAQ. Scientists who helped us, including young STS scholars such as laboratory study researchers, were well aware of the bias scientists speaking out are placed under during emergencies (e.g. Minamana disease, Chelnobyl or Hurricane Katrina). While putting the FAQ together, we tried to lighten up any sensationalized, or strongly advocated comments in deficit-model manner, and maintained a perspective that reflected the scientific evidence available.

As a result, visitor numbers to the FAQ shot up in record numbers. Within half a month, the page had topped more than 1.3 million views, and with the help of volunteers, the page was translated into English, Mandarin, and Spanish.[9,10]

However, looking back on the FAQ today reveals many unfortunate mistakes. We cannot say the page never tried to make people feel safe, and it has been criticized as "implying that the situation was far less severe than what it really was." In a way, this is true. But in such chaotic times, there is a limit to how much one can communicate. Whenever with the understanding how essences of misunderstanding is misunderstood between experts and lay peoples in such cases (cf. Wynne 1996), perhaps this is the limit to what we could offer.

3.5.3 Sending Out Science Alerts

Science Alerts (SA) are a priority to the SMCJ, and they are sent out throughout the year. By looking at the titles of the SAs the SMCJ produced following the Great East Japan Earthquake (Table 3.1), it is possible to get a glimpse into the issues the

---

[9] An accurate record of viewers is unavailable due to the fact the SMCJ server crashed several times because of internet traffic to the SMCJ website. We are grateful for our server company who then provided us with free server rental and unlimited bandwicth during this time, a story that was also picked up by the media.

[10] Some complications were experienced here. When the SMCJ server crashed, the University of Tokyo graduate students who we had been collaborating with had created and uploaded their own SMCJ website and continued to update information through this website. While their intentions were good, as an officially registered independent organization we could not allow them to use the SMCJ name without our consent. But more importantly, it was the risk the students learning about atomic power were putting themselves and the public under by uploading information about health safety. Within the day, both groups negotiated and agreed to transfer content to the SMCJ. Later, our collaborators would be silenced by a gag order given out by the University of Tokyo.

Table 3.1 SAs sent out by the SMCJ

Date[a]	Title	# of experts	Comments	Tweets	Bookmark
15-Mar-11	An expert's estimate about the events at Fukushima Daiichi Nuclear Power Plant	1	7	8	78
16-Mar-11	Experts reaction on Japan nuclear incidents (F)	2	0	0	2
16-Mar-11	Q&A on the Nuclear Power Station	–	–	1,850	555
18-Mar-11	An expert's reaction about radiation exposure	1	3	1	5
18-Mar-11	Rolling blackouts, are they valid or not?	1	6	11	1
18-Mar-11	Expert response to dispersal of radioactive material	1	4	1	0
18-Mar-11	Expert response to radiation exposure	1	8	24	0
19-Mar-11	Experts on nuclear power station and radioactive materials (F)	7	0	2	0
19-Mar-11	Experts' comment about the Tohoku earthquake	2	0	0	0
20-Mar-11	Links to sources of information about Tohoku earthquake	–	18	229	20
20-Mar-11	Possibility of recriticality	1	23	20	0
22-Mar-11	Effects of radioactive materials	1	57	322	27
23-Mar-11	Radioactive iodine in tap water (F)	1	0	0	0
26-Mar-11	Filtration of radioactive materials by water cleaner	1	9	6	0
28-Mar-11	How to estimate personal exposure dose	1	4	36	12
29-Mar-11	Effects of large earthquakes on a global scale (F)	1	0	10	0
1-Apr-11	Contribution: comparing fallout and current situation	1	39	159	50
1-Apr-11	Prerequisite of risk communication	1	11	161	7
1-Apr-11	An expert's reaction about internal exposure	1	22	251	59
2-Apr-11	Experts' reaction about radioactive food contamination- SMC Canada (F)	–	0	4	0
5-Apr-11	Human impact of low dose exposure	1	22	1,911	365
5-Apr-11	Decreasing amount of coral in waters close to Japan	5	0	0	4

(continued)

Table 3.1 (continued)

Date[a]	Title	# of experts	Comments	Tweets	Bookmark
7-Apr-11	Expert reaction: stem cells generate eye-like structure	2	0	8	2
8-Apr-11	Directional movement about a nuclear energy policy after 3.11	1	20	837	165
12-Apr-11	Expert reaction to the raised severity rating at Fukushima (F)	3	0	239	46
17-Apr-11	There is a crack in everything- radioactive contaminants' from "food, sex, and salmonella: why our food is making us sick" by Dr. David Waltner-Toews.	–	0	351	54
22-Apr-11	About phytoremediation	1	0	128	15
13-May-11	Experts reaction: artificial feeding the bear	3	0	187	17
29-May-11	The eruption of mount Merapi in Indonesia	2	0	2	1
30-May-11	Experts' reaction about recommendations of ICRP and ECRR	2	0	763	131
31-May-11	Risk of food poisoning in the case of blackouts	1	0	44	10
11-Jun-11	Contribution: disaster area and its governance	1	0	30	6
11-Jun-11	"Genetic fingerprint" from low dose exposure and its application potency	2	0	64	3
21-Jun-11	Cancer risks attributable to low doses of ionizing radiation: Assessing what we really know' (Brenner et al. 2003) translated by Dr. Masashi Shirabe	–	0	2,872	279
21-Jun-11	Dr. Masashi Shirabe's comment to Brenner's article	–	0	579	219

[a] Date indicates final update
F comments provided from other SMCs

SMCJ wanted to contribute towards. Normally, SAs are sent to journalists, and then uploaded to the SMCJ website a few days later, but an exception was made during the disaster where SAs were sent to journalists and uploaded to the website at the same time. We also unlocked the comment section to make it possible for the public to give us feedback.

However, looking at the list of SAs also reveals a gap between the ideal and reality. The number of SAs themselves only number 35 between March and June 2011, and only 10 of them contain comments from several experts. By comparing

SAs related to the disaster to SAs not related to the disaster, which were both released around the same time, the number of expert comments was often one against several. It was clear researchers being asked about the nuclear accident were keeping their distance from the political and social debate. While similar numbers of experts were approached for comments in both cases, it was difficult to find experts willing to comment on the accident.

A significant issue is that these researchers, who until the accident had been publishing research papers on this complex problem, were employed by organisations who chose the Unified Voice option immediately following the accident. That is, the only answer researchers could give people was, "please wait until my organisation releases its statement".

This was in contrast to the SMCJ's approach in an emergency, which is to release expert comments as they came out. Using the SMCJ website's comments section we were able to monitor reactions from new experts, a number of which we followed up by negotiating with the expert to receive a more in depth comment on an issue. These additional comments were also released and uploaded to the website. But this took time, and it highlights the challenge of collecting a variety of expert comments on one complicated issue.

A valuable expert source during the disaster was the SMC's global network, where SMCs overseas could share comments they had collected from experts in their own countries. This proved important after it was announced the nuclear power plant accident would be raised to a Level 5 accident. While the raise was expected, we also predicted it would make it even more difficult to find a Japanese expert willing to comment on the scientific reasoning behind the raise. Most experts were either already working with government officials, or they were at risk of being labeled a scientists who defends government agendas. The SMC global network helped provide the SMCJ with a list of experts who could provide a comment, and so by the time the accident level was raised again to Level 7, the SMCJ was already ready to collect and send out comments from overseas experts.

Yet looking back on the SAs themselves reveal an imbalance in views. This discourse was most likely caused by the normative bias displayed by scientists in a crisis to use science to calm the public. The way in which the SMCJ overcame imbalance was to open up the comment section below each SA. Conversations between skeptic members of the public and experts lead to deeper discussions, and effectively the comments section acted as an antidote for imbalance.

3.5.4 Other Activities and Independence

Many other activities the SMCJ had carried out also deserve to be mentioned here.

Firstly, an online links page developed by the SMCJ was a useful information source for people looking for information related to disasters and medical assistance.

The page, "Links: Disaster Relief Information for Vulnerable Citizens[11]" consisted of a collection of links to pages developed by various organizations on topics ranging from where to find specialized medical facilities, to disaster victim support. The page was developed together with the help of a medical sociology and science in society expert at Kanazawa University, and was also released as a text document.

The SMCJ also received a number of contributor articles. The most significant includes a research paper published in the Proceedings of the National Academy of Sciences, "Cancer risks attributable to low doses of ionizing radiation: Assessing what we really know (Brenner et al. 2003)", which was translated into Japanese by STS expert Masashi Shirabe with the author's permission. The paper was subsequently referred to in a number of debates concerning low dose radiation exposure.

Another article included an article reviewing the effects the Chernobyl accident had on local food produce and the environment by Canadian epidemiologist David Waltner-Toews (Waltner-Toews 2008). With permission from Professor Waltner-Toews, who was introduced to the SMCJ through the SMC-Canada, the article was translated into Japanese and used to predict the effects radiation would have on farm produce surrounding the Fukushima Dai-ichi Nuclear Power Plant.[12]

The SMCJ also hosted out two Lesson Learning discussions both in Tokyo and in Fukushima. The 54 who participated in the Tokyo event, and 42 who participated in the Fukushima event comprised of scientists, journalists, bureaucrats and local citizens, who were invited to talk about their views on the relationship between science and the media following the Great East Japan Earthquake. Under the Chatham House Rules, participants reflected on decisions they regretted, and engaged in discussions about what to do next. Both events were an opportunity for different stakeholders to meet and listen to different perspectives of the disaster, and has contributed to several policies, and helped give Fukushima a voice in the national media to issues they thought were being overlooked (SMCJ 2014).

To make it clear, the SMCJ has worked to maintain its independency from the time it was established until today. Gratefully, government and major companies have agreed it would be better this way. On March 16, 2011, the SMCJ received a phone call from the Cabinet Office. Initially fearing we were going to be asked to remove SAs with controversial comments from scientists, to our surprise, they simply asked us whether there was anything they could do to help.[13]

The SMCJ was awarded for all of its efforts by National Institute of Science and Technology Policy. While we are glad the SMC movement has been acknowledged as an important resource, it has also been subject to a fair amount of criticism. This

---

[11] http://smc-japan.org/?p=845.

[12] For example, Fukushima is a region famous for producing shiitake mushrooms, and this was the first food source to be affected.

[13] In this connection, our answer for this question was a request for English speakers. At this time we were very busy with corresponding with the Japanese media and had no energy to respond to enquiries from oversea. But their response was that government also had a shortage of English speakers.

includes "putting the public in danger by releasing comments during times of uncertainty", and "controlling the science agenda in the media".

3.6 SMCJ Activity Following the 2011 Great East Japan Earthquake: A Critical Look

The Science Media Centre of Japan (SMCJ) is a movement that has experimented with introducing a different approach to Japanese media's agenda setting architecture. Let us critically examine the way in which the SMCJ used its agenda building approach for the media during the Great East Japan Earthquake. Of course, studying the relationship between the Fukushima Dai-ichi Nuclear Power Plant accident and the media is not easy, and it is difficult to depict an overall view. However, the way in which communication broke down between experts and society is a variation of what has happened in past disasters such as Hurricane Katrina, with the exception that this was a disaster comprised of several elements, making it that much more complex.

What is clear is that the scientific community remained silent. Not because the media was forcing an "everything is safe" agenda on them, nor because of the Japanese culture. As a matter of fact, months later many experts confessed they knew the that telling the public everything was safe was not the right solution, but they did not raise their voices because they thought "what difference would one person make against an entire tide (Independent Investigation Commission on the Fukushima Daiichi Nuclear Accident 2012; 7)". In other words, "we had a situation where several scientists had the same view on one issue but still remained silent (Kitamura 2012; 723)". This lack of communication between scientists revealed a social system failure within the scientific community.

The SMCJ chose to tackle the issue face on. We may have only thrown a small stone into the large sea of media chaos following the earthquake, but we believe it is well worth documenting our activities to identify and examine any previously unknown challenges.

Before we jump into anything else, it is important to make the limits of the SMCJ activities clear. The SMCJ's goal is to provide an alternative method for setting the social agenda, which today is still very much controlled by an elite group of scientists and journalists. Our fundamental architecture does not place us in between the public and experience experts. It is its own immanent limit. Nonetheless, the SMCJ does make an effort to extrapolate the public's perspective using STS research.

With this in mind, this section will start with who and how the SMCJ selected experts to talk following the earthquake, then move onto the development of new platforms such as social networks which the SMCJ now uses, and finally what issues the SMCJ currently faces, and what it plans to do next.

3.6.1 The Agony Involved in Finding an Expert

The SMCJ has repeatedly sent out Science Alerts to journalists, which are made up of several expert comments concerning one scientific issue. An important part of the Science Alert process is deducing who is the best expert to ask for a comment. Not only that, but the expert must also be willing to pass their voice onto the media through the SMCJ. So while a total of more than 100 experts helped the SMCJ, several hundred experts had already rejected us. As a result, we have found that the best experts are often the ones who are willing to engage.

The most important question members of the SMCJ discuss about before contacting an expert is how we maintain balance, and what questions do we ask. This is particularly important since experts are more likely to give different views on risk during a time of uncertainty (cf. Bruine de Bruin et al. 2006; Kosugi et al. 2011). Sometimes, it turns out an expert specializing in a field surrounding the issue in question has a far clearer perspective on risk assessment than an expert directly involved.

We managed to build a process to collect expert comments, but that does not mean the comments are unbiased since it is the SMCJ that puts together the Science Alert.

Encouraging experts to comment during a crisis is difficult. The organization scientists work for often see protecting the organization's social status as its biggest priority, and it is common for them to opt for the Unified Voice option. As a matter of fact, faculty staff and graduate students at the University of Tokyo were warned not to speak to the media. Enterprises such as the National Institute for Radiological Sciences released statements on behalf of all its scientists. These components lead to the silencing of scientific community within Japan, which is a problem for groups such as the SMCJ who are trying to look for experts expressing the majority of the community.

Minority comments were also collected occasionally, but that does not mean the final Science Alert contains 50 % majority comments and 50 % minority comments. Science Alerts provide a look into the weight of opinions throughout the scientific community. This means at times, the minority voice could be overpowered by other comments.

In general, the easiest comments to obtain are from those in the minority group who have strong motivations towards one idea. But these comments are often sensationalized. As Boyce stated in the MMR vaccination scandal, such comments can only create an over-balance in opinions (Boyce 2007).

The challenge the SMCJ faced following the Great East Japan Earthquake was whether the critical comments the team had been collecting reflected the opinion of those across the entire scientific community. Immediately following the earthquake, the SMC team members agreed the important issue was providing information on the nuclear reactors, and a debate followed discussing how to find experts. Most experts had been summoned to the Prime Minister's Office, while others were locked inside their respective academic associations. Finally, two names came up of experts who had participated in a consensus meeting in 2000 regarding the geological disposal of nuclear waste.

One expert worked at a university as an atomic energy expert, but was also famous for being an opinion leader against nuclear power plants. The expert had already started writing articles for anti-nuclear power plant citizen groups immediately after the earthquake. The other expert was an atomic energy researcher considered to be a key person within Japan's Nuclear Power Village, and someone who had already experienced the sensitivity of their field inside society.

After some heated discussions, the team decided to approach the second expert first, seeing as the media had not interviewed them yet. The first expert had already been answering questions from the media, and so we believed it might not be too difficult to ask for his help. Also, the danger of only relying on this expert's comments was that the SMCJ risked being labeled an anti-nuclear organization. After a long debate, it was decided that we would not use this expert as an expert commentator without the second expert. But we also never heard back from the other scientist because they were occupied with reading and answering questions from the general public.[14] To summarize, the SMC's idea to find an expert to comment on an announcement ended in failure. In other words, the SMCJ feared being labeled as a group with certain ideologies, but our attempt to seek balance, ended in failure.

It is difficult to measure the impact of this idea. To date the SMCJ staff is divided on whether to accept this failure as it is, or whether there was something else we could have done. Amongst them is failing to produce a worst-case scenario.

However, when it comes to thinking about what should have been done, the answer still remains unclear. Considering the fact that the situation on site during the crisis was one far beyond any predicted catastrophe, it was difficult to find scientists who could speak with an open mind. For the media, their priority was to get the political view on the situation, and then seek experts' advice (Wien and Elmelund-Præstekær 2009). "Science, if it can deliver truth, cannot deliver it at the speed of politics (Collins and Evans 2007)." SMCJ's trial to rebelling for this tide is also inevitably tinged with political color. In fact, every organization's trials to grasp the scientific situation in a catastrophic atmosphere directly meant assertion of their own ideology. Organizations with various ideologies were trying to use experts politically. During the SMCJ's activities, we asked for comments on experts appearing on the liberal press or anti-nuclear groups, but those organizations kept their experts trapped up. Of course, the best experts were ushered to the Prime Minister's office. Every organization took action to enclose their experts.

Just after the disaster, it looked like individuals were having calm discussions in places such as SNS because they were in a disaster utopia (Solnit 2009). However,

---

[14] As a result, the comments provided by this expert within the Nuclear Power Village was picked up by one of the SMCJ's registered journalists, and used in a tabloid news site. Even though the article was accurate, they were unable to make a difference in the upstream engagement process at bigger newspapers.

between organizations, there were ideological conflicts. The SMCJ tried to mediate them, but ended up becoming part of the conflict.

Moreover, experts' comments from SMCs in other countries were welcomed in Japanese society as the neutral opinion, but we must reconsider this. Of course, we tried to add information such as career, position or conflict of interest, but Japanese journalists and audiences did not have enough knowledge about such context. There is no denying that the Japanese were expecting excessive objectivity from those comments.

3.6.2 When Scientists Begin to Protect Themselves

Described in the last section were difficulties on taking experts' comments, but it also hints at the difficulty of constructing the "*suitable* worst case scenario" in the open forum of the media. Moreover, that is a problem concerning *who* becomes the representative that talks about this scenario to the public. To make a deliberative forum, or treating a scenario as it has fallibility, the representative is usually an individual person, not an organization. It is common for organizations to promote a unified message of its constituent members.

Undertaking the role of representative in a crisis situation with maximized uncertainty and ambiguity means becoming a direct target of credibility. Because of its norm as a scientist or citizen, experts are bound between responsibility for their community and accountability for the public. Especially in Japan, object of credibility had been the organization's system, and this system was functioning fine until the disaster. In such circumstances, the danger of undertaking the role of representative is shared between experts.

The SMCJ tried to intervene with this circumstance, and tried to encourage experts to come out and approach the media in order to accomplish public accountability. But for the audience, every actor on a stage is an unreliable narrator. This fact became clear day by day after the disaster, and the disaster utopia gradually disappeared. Protecting experts became a new task for the SMCJ.

Enquiries (query) left on the access logs of SMCJ website indicate interesting tendencies. Just after the disaster, scientific terms such as "nuclear" or "radioactive" ranked high on terms used in the search engine. However, about a month after the earthquake, netizens started visiting the SMCJ website to look for particular experts. In other words, in the first few days people wanted to get a grasp on the situation scientifically and visited our website. But later, they started looking for human targets of trust or distrust. At this point, comments from visitors started to show dichotomy. Namely, they sought their enemy and left criticism, or sought their hero and left comments of praise.

In this latter stage, expert comments in comparatively older entries on the website became a target for criticism. For example, after the disaster the Kanto region started rolling blackouts. At first, there was a slight reaction to expert comments about the measure. For example, estimates of supply and demand of electric power, But later, people started criticizing individual experts in the comments claiming he could be a pawn owned by TEPCO sent to legitimize the need for nuclear power. Indeed, it is difficult to say all experts are free from the political economy of such enterprises. This fact complicated the argument.

Under this pressure, expert comments became longer and longer, and started including verbose contingent conditions. However it must be noted that these lengthy contingent conditions were not only to maintain scientific accuracy, but also were reflecting experts' image of the audience. In other words, they understood that the situation couldn't be resolved only by science. Sometimes experts refused to comment because of the consciousness of their subjective point of view. For example, the SMCJ asked several experts to make a paper review about issues confronting Japan, but all of the experts denied this request. Some answered that it is a task for academic societies, and others said "reviews always contain my subjective point of view". Refusing to comments does not just mean abandoning responsibility. Sometimes denial is a reflection of understanding society, or integrity as a citizen.

The safest choice for experts in the catastrophic situation was to keep a distance from the argument. Therefore, every experts talked to public must be regarded. The SMCJ must have a function to protect experts who stepped out from the hives of their peer communities. Without this function, diversity of the opinion from science community cannot be maintained. This is one task for us.

3.7 Realization of Agenda Building: Spiral of Silence and Two Scientists

To conclude this chapter, we will discuss the relationship with social network sites (SNSs). The SMCJ's goal has been to reflect and construct diverse opinion from the science community to the media, and bringing the minority into the science debate. Talking about this point of view after the disaster, SNSs cannot be dismissed. SNS exposed expert voices directly, and it is a place where agenda is directly constructed. With expectation of these functions, the SMCJ also engaged on SNS and picked up on some agendas from this new forum. In fact, some of the agendas the SMCJ adopted into Science Alert originated from lay person or expertized netizens on SNS.

However, can agendas formed on SNS be reliable, and do deliberations consist of diverse opinions? According to our study, content diversity on socio-scientific issues were high in newspapers, portal sites, and blogs in total, but surprisingly low on Twitter (Shineha and Tanaka 2014; Tanaka et al. 2012). Social media seems not to be public sphere, but seems to act like community generator. Even today, socio-scientific agendas and opinions are arranged on main stream media (press, televisions, etc.), and their salience is catalyzed on SNS by the credibility network. With barely legitimized justice on their back, netizens chose the object of trust, and tied together similar acts of trust, which finally ended in the formation of exclusive communities.[15] This community is an echo-chamber (Sunstein 2001) for individuals, and the peripherals of the community act like a filter bubble (Pariser 2012). That is, a place to resonate each ideology, complete with a membrane that only permeates information that unites the community. We should consider more about the empirical fact that malicious suppositions on network edge often produce a psychological gap between nodes with preservation of the tie. Lessons we learned from the disaster was that SNS could be a device for networking, but it could also curtail diversity. From the point of view of risk communication, this architecture is useful during the care communication stage when elite indexing (Bennett et al. 2007) is welcomed, but sometimes it is harmful in the stage of consensus communication regarding discussions about the distribution of risk in No Spin Zone (Bennett et al. 2007).

Looking at these functions, SNS in general provides a message platform for experts willing to fulfill their accountability, and that every scientific message has the possibility to be 'flamed'. The problem is that this repulsion can raise the fear of isolation, and atmosphere of these feelings can cause a 'spiral of silence (Noelle-Neumann 1984) . Under such conditions, the selection of experts occurs: someone who advocate along populism, or someone who can endure the flames against his or her discourse are the survivors. Ironically, these remaining experts are recognized as trustworthy experts among their followers.

The SMCJ tried to bridge the gap between dichotomies by operating on an upstream of information in the manner of science, and by shedding light on minority groups in scientific arguments. But, can an artifact like SMCJ spin critical discourse toward hegemonic perspectives by mashing up expert comments? Related to this question, Tateishi (2014) is examining papers published after the disaster, and indicated a tentative assumption. He cites 6 of the following conditions that criticism can be consistent with scientific research: (1) Research is not in 'big science' stage, (2) Researchers have access to actual issues, (3) Openness of the outcome of official survey, (4) Embedded in international and trans-disciplinary academic community, (5) Academy has capacity to evaluate historical or local nature of science objects such as the atomic bomb or the Chernobyl accident, (6) have a way to reflect social consciousness into research outcome.

These are assumptions about paper production, but most of these conditions coincide with difficulties that the SMCJ faced after the disaster. An expert accord to some of the above conditions seems easier to talk about to the public through media, and vice versa. To put it the other way around, the task of the SMCJ is how we can collect comments or help experts talk to the media to accomplish their accountabilities Adding to the condition from an empirical fact from active experts on SNS after the disaster that is 'locating in quasi-expertise field'. Experts who

---

[15] After the disaster, targets for credibility had focused on SNS, and this focusing induced divide of the community on SNS. We are now trying to depict this process with social network analysis of Twitter data.

became intermediate information centers or hubs for discussion on SNS were not specialists in fields like nuclear power plant engineering or radiology, but from adjoining fields such as nuclear physics, high-energy physics or computer science. These quasi-experts were free from direct political economy of the issue, and therefore they could inspect the problem as an independent scientist. They maintained a canonical view of science throughout communications with other experts (including exact experts) and concerned citizens, they showed interactional expertise (Collins and Evans 2007) and succeeded in care communication. A lay tweet from followers of those expert discussions may be an evaluation for interactional expertise: "Researchers related to nuclear power are biased a lot of the time, so I do not rely on them. Natural science researchers who studied nuclear power in an improvisional manner after 3/11 are less noisy and neutral."[16]

However, despite the presence of such quasi-experts, difficulty remains in achieving agenda building through ad hoc consensus communication. Two contrasting heroic experts on SNS represent this difficulty. One of them, astronomer Professor Junichiro Makino is considered to be the person who most accurately anticipated the situation after the disaster. Using his own blog and twitter account, he calmly continued analyzing available data and keep questioning the conservative atmosphere in society following the disaster. But it must also be stated that his scientific point of view does not reflect the majority. Makino, who also has a profound knowledge of STS studies, kept analyzing data in a precautionary approach, but that may be because of his consistent attitude that science has fallibility (cf. Makino 2013). Sometimes his discourses sound cynical in the SNS sphere and he remains in minor.

There is no doubt among current Japanese media atmosphere that Professor Ryugo Hayano is another hero from SNS. In contrast to Makino, he had been an important hub for information, and also been interactional expert for trans-disciplinary studies. These two experts actions and influence on SNS are thought-provoking. Hayano says "refutation on the Internet causes an exchange of negative words (Fukushima Minyu 2014)". However, whether he likes it or not, his discourse sublimates into hegemony of scientific safety in a socially constructive manner. When he acts as an (interactional) expert and he published a paper indicating that current conditions of irradiation will not cause severe harm (cf. Hayano et al. 2013), and retweeted the article introducing his study, his advocacy was amplified explosively by his 120,000 followers. That is a way for a interactional expert in the social network age.

From the point of view of this book, it can be judged that Hayano has already been a politically-constructed symbol. But actually, the majority of reputation in Japanese social media is that Makino is too political and STS-like,[17] while Hayano is the scientist. This is the dilemma where the SMCJ is heading. Practically, as

---

[16] Hosuke Nojiri, @nojiri_h a science fiction writer, "(tweet fully quoted in body)," 3 July 2011, 17:33. Tweet.

[17] After the complicated argument, representation of the word 'STS' was sometimes used for ridicule in Japan after the disaster, especially on twitter.

Makino showed his model, the SMCJ is trying to adjust the balance of unfair distribution of risk precautions by helping contribute to the healthy and diverse scientific argument. Simultaneously, the SMCJ is also trying to be a hub for scientific information and to obtain effective power on agenda building just as Hayano did. To fulfill this goal, the SMCJ must challenge to subsume these two different types of experts, and also must reflect the perspective of lay experts. Companying with this dilemma is one of the major tasks for the SMCJ.

3.8 Building the Agenda to Bridge the Gap

Let me change to the summary by presenting a single article. Three years after the disaster, Fukushima Minyu, a local newspaper in Fukushima published a serial of 100 articles that got to the heart of the issues. They are *blended-commentary* styled articles, where both experts and lay experts are mentioned in the same article. According to the journalist at the Fukushima Minyu, after three years the atmosphere in Fukushima had calmed down enough to write about such friction between residents. This serial eloquently depicted the actual conditions of socio-scientific issues, and also demonstrated the difficulties of framing the issue with agendas. Professor Ryugo Hayano appears in an article of the serial titled "Distrusted safety: Rejecting the data gathered by researchers":

> Hayano insisted that personal dosimeter could measure the exposure based on the actual life of every residents. The government also introduced it in released evacuation areas. But Taguchi (resident, leader of civic group) is annoyed that personal dosimeter shows lower exposure dosage than the conventional uniform calculation method of exposure that presumes every resident had spent 8 hours outside. With personal dosimeters, dosages vary between each individual in the same household, depends on how each individual lived every day.
>
> Taguchi said that, "measuring personal dosage that reflects actual lifestyle may be right," but he can not stop being suspicious. "By changing the dose management from air dose to personal dose, is the government trying to divide the residents?".
> (Fukushima Minyu March 4 2014)

Indeed, from a sociological perspective, when unified risk management is distributed to individuals, whether the government intends it or not, there is a possibility of divide. Is it possible to answer Taguchi's question by scientifically reframing the problem? To relieve the power of the number, what kind of expert should SMCJ ask for comment? It is not easy to solve, but worth tackling problem.

3.9 For the Future

In this chapter, I tried to depict the location and activity of SMCJ in the Japanese science-media landscape and tried to grasp the reflexive meaning of the Great East Japan Earthquake.

But the biggest problem still remains. The SMCJ's attempt to invite experts out to meet the media and encourage them to accomplish public accountability. During this activity, the SMCJ's power on agenda setting is hidden behind the scene. It is the nature of intervening artifact, but we need deep reflection and strict norms on this nature. This problem must continue to be discussed.

In the end, it is needless to say that this article is a reflection of my own perspective. The other staff in the SMCJ, and SMCs in other countries may have different opinions. Intervening the process of legitimization of socio-scientific risk is an act full of conflict. I hope this manuscript may have some significance for further discussion.

Acknowledgments If SMCJ's activity after the disaster was worthy of great praise it received, it is thanks to the dedicated staff and supporters who endured those severe days. I would like to show my greatest appreciation to Miho Namba and Motoko Kakubayashi, the manager and international officer at the SMCJ at that time. Also, I am also in great debt to other supporters including Akane Shirota, Makiko Watanabe, Daisuke Yoshinaga, Kentaro Nagai, Mitsuru Kudo and many persons who joined our program. And finally, I am in debt to Professor Shiro Segawa for his guardianship to our project.

References

Asayama, S., & Ishii, A. (2012). Shimbun houdou ni yoru IPCC zou no kouchiku to sono shakaitekigani (Framing analysis of climate science in Japan: An analysis of Japanese major newspapers' reporting on the IPCC). *Kagakugijutsu Shakairon Kenkyu, 9*, 70–83.

Bennett, W. L., Lawrence, R., & Livingston, S. (2007). *When the press fails: Political power and the news media from Iraq to Katrina*. London, UK: University of Chicago Press.

Boyce, T. (2007). *Health, risk and news: The MMR vaccine and the media*. New York: Peter Lang.

Brenner, D. J., Doll, R., Goodhead, D. T., Hall, E. J., Land, C. E., Little, J. B., et al. (2003). Cancer risks attributable to low doses of ionizing radiation: Assessing what we really know. *PNAS, 100*, 13761–13766.

Bruine de Bruin, W., Fischhoff, L., Brilliant, L., & Caruso, D. (2006). Expert judgments of pandemic influenza. *Global Public Health, 1*, 178–193.

Bruns, A. (2008). *Blogs, wikipedia, second life, and beyond: From production to produsage*. New York: Peter Lang.

Collins, H., & Evans, H. (2007). *Rethinking expertise*. Chicago: The University of Chicago Press.

Davis, N., (2008). *Flat earth news*, London, UK: Random House.

Fukushima Minyu. (2014, March). Genpatsu saigai fukkou no kage–shinrai sarenai "anzen" kenkyusya ga atumeta data ni "kyohihannou" (Distrusted 'safety': 'Rejection' toward the data gathered by researchers). Minyu-net. Retrieved July 5, 2014, from http://www.minyu-net.com/osusume/daisinsai/serial/fukkou-kage/140304/news.html. Electric version. Serial of articles are available from http://www.minyu-net.com/osusume/daisinsai/serial/fukkou-kage/index.html.

Gaver, I. (2001). Government by spin: An analysis of the process. *Media, Culture and Society, 22*, 507–518.

Hallin, D. C., & Mancini, P. (2004). *Comparing media systems: Three models of media and politics*. New York: Cambridge University Press.

Hayano, R. S., Tsubokura, M., Miyazaki, M., Satou, H., Sato, K., Masaki, S., et al. (2013). Internal radiocesium contamination of adults and children in Fukushima 7–20 months after the Fukushima NPP accident as measured by extensive whole-body-counter surveys. *Proceedings of the Japan Academy. Series B, Physical and Biological Sciences, B89*, 157–163.

Hakuhodo DY Media Partners. (2011, October). Shinsai no eikyou ni yoru seikatsusya no media sessyoku jyoukyou ni kansuru chousa (Survey about media usage of citizens after the disaster, Trans.). Survey report. Retrieved May 30, 2014, from http://www.hakuhodody-media.co.jp/wordpress/wp-content/uploads/2011/10/HDYnews111031.pdf.

Hayashi, K. (2011). *<Onna Kodomo> no Journalism–Care no rinri to tomoni (Journalism of women and children: with ethics of care)*. Tokyo: Iwanami Shoten.

Independent Investigation Commission on the Fukushima Daiichi Nuclear Accident. (2012). *Fukushima genpatsu jiko dokuritsu kenshyou iinkai chyousa kensyou houkokusyo (Independent investigation commission on the Fukushima Daiichi nuclear accident, final report, Trans.)*. Tokyo: Discover 21.

Kageura, K. (2013). *Shinrai no jouken–Genpatsu jiko wo meguru kotoba (Conditions of trust: Discourses around nuclear powerplant disaster, Trans.)*. Tokyo: Iwanami Shoten.

Kitamura, M. (2012). Genshiryoku anzen ronri no saikouchiku to resilience base no anzengaku (Reframing of nuclear safety logic on the basis of resilience engineering). *Nihon Genshiryoku Gakkaishi, 54*, 721–726.

Kosugi, M., Tsuchya, T., & Taniguchi, T. (2011). *Gijutsu risk ni taisuru senmonka to simin no shiten: ippan shimin tono kairi wo kanjiru senmonka no tokutyou (Viewpoints of the "expert" versus "the public" on technological risks: Characteristics of the expert with a feeling of distance from the public)*. *Nihon Risk Kenkyu Gakkaishi, 21*, 115–123.

Lang, G. E., & Lang, K. (1983). *The battle for public opinion: The president, the press, and public opinion*. New York: Columbia University Press.

Lewis, J., Williams, A., & Franklin, B. (2008). "A compromised fourth estate? UK news journalism, public relations and news sources". *Journalism Studies, 9*(1), 1–20.

Lundgren, R. E., & McMakin, A. H. (2011). *Risk communication: A handbook for communicating environmental, safety, and health risks* (4th ed.). Hoboken: Wiley-IEEE Press.

Macilwain, C. (2012). Two nations divided by a common purpose. *Nature, 483*, 247.

Makino, J. (2013). *Genpatsu jiko to kagakuteki houhou (Nuclear power plant accident and science methods, Trans.)*. Tokyo: Iwanami Shoten.

McCurry, J. (2014, May). Gourmet manga stirs up storm after linking Fukushima to nosebleeds: Outcry over storyline blaming radiation exposure moves PM to respond and publisher to suspend Oishinbo series. *The guardian*. Retrieved May 23, 2014, from http://www.theguardian.com/world/2014/may/22/gourmet-oishinbo-manga-link-fukushima-radiation-nosebleeds.

MEXT (Ministry of Education, Culture, Sports, Science and Technology—Japan). (2012). FY2012 white paper on education, culture, sports, science and technology (English version). Retrieved April 23, 2014, from http://www.mext.go.jp/component/english/__icsFiles/afieldfile/2013/01/15/1329765_05.pdf.

Nagai, K. (2015). Chikyu ondanka heno media attention (Media attention of global warming to politics than science, Trans.). In S. Segawa & N. Sekiya (Eds.), *Henyousuru Kankyomondai to Journalism (Transfiguration of environmental issues and journalism, Trans.)*. Kyoto: Minerva Shobo. (Chapter 6).

Namba, M., Tanaka, M., & Saijo, M. (2014). *Providing of scientific information in the nuclear accident: Settle on Fukushima Daiichi nuclear plant accident after 2011 Tohoku earthquake*. Paper presented at the 6th International Conference on Knowledge Management and Information Sharing, Rome.

NAOJ (National Astronomical Observatory of Japan), & Sokendai (The Graduate University for Advanced Studies). (2005). Gakujutsu Seika no Symposium (Public relations and press about scientific research). (Report).
Noelle-Neumann, E. (1984). *The spiral of silence: Public opinion—Our social skin*. Chicago: University of Chicago.
Ooi, S., Ogawa, K., Kobayashi, Y., Sakoh, S., Fukuda, M., Yamamoto, K., et al. (2014). *2013 nen ban "Nihon no journalist chousa" wo yomu—Nihon no journalism no genzai*. (Reading journalism survey in Japan (2013 edition): Present stage of Japanese journalism, Trans.). *Journalism and Media, 7*, 247–279.
Office of Science and Technology, & The Wellcome Trust. (2000). Science and the public: A review of science communication and public attitude to science in Britain. London: The Wellcome Trust.
Pantti, M., Wahl-Jorgensen, K., & Cottle, S. (2012). *Disasters and the media*. Oxford: Peter Lang.
Parisar, E. (2012). *The filter bubble: How the new personalized web is changing what we read and how we think*. London: Penguin Books.
Pinholster, G., & O'Malley, C. (2006). EurekAlert! survey confirms challenges for science communicators in the post-print era. *Journal of Science Communication, 5*, 1–11.
Protess, D. L., & McCombs, M. (1991). *Agenda setting—Readings on media, public opinion, and policymaking*. NJ: Lawrence Erlbaum.
Qi, H. (2014). *Shimbun houdou ni oite housyanou no kenkoueikyou risk wo kouseisuru actor gensetsu no bunseki* (Discourse analysis of actors constructing radioactive risk on newspapers, Trans.). Dissertation, Waseda University, Tokyo.
Royal Society. (1985). *The public understanding of science*. The Royal Society.
Reporters Without Borders. (2014). 2013 World Press freedom index: Dashed hopes after spring. Retrieved April 22, 2014, from http://en.rsf.org/spip.php?page=classement&id_rubrique=1054.
Segawa, S. (2010). Kenkyusha no mass media literacy (Mass media literacy of science experts). JST-RISTEX project survey report.
Segawa, S. (2011). *Genpatsu houdou wa 'daihoneihappyou' dattaka –asa, mai, yomi, nikkei no kijikara saguru* (Was nuclear reactor report 'imperial headquarters' announcement?': Comparison from Asahi, Mainichi, Yomiuri and Nikkei articles, Trans.). *Journalism, 255*, 28–39.
Séralini, G. E., Clair, E., Mesnage, R., Gress, S., Defarge, N., & Malatesta, M. (2012). Long term toxicity of a roundup herbicide and a roundup-tolerant genetically modified maize. *Food and Chemical Toxicology, 50*, 4221–4231. (Retracted).
Shineha, R., & Tanaka, M. (2014) "Mind the gap: 3.11 and the information vulnerable" *The Asia-Pacific Journal, 12*(7), 4.
Shineha, R., Yashiro, Y., & Tanaka, M. (2014). *Analysis of media discourses on stem cell research and regenerative medicine in Japanese newspapers*. Paper presented at International Society for Stem Cell Research 12th Annual Meeting, June 18–21, 2014, Vancouver, Canada.
SMCJ (Science Media Centre of Japan). (2014). "Lessons learnt after 3.11," summarize of discussion in Tokyo (Sept 11, 2011) and Fukushima (July 27, 2012).
Solnit, R. (2009). *A paradise built in hell: The extraordinary communities that arise in disaster*. New York: The Viking Press.
Sunstein, C. (2001). *Republic.com*. New Jersey: Princeton University Press.
Tanaka, M., Shineha, R. & Maruyama, K. (2012). *Saigai Jyakusha to Jouhou Jyakusya -3.11 go naniga misugosaretaka?* (Vulnerable populations in disaster and digital divide: What was ignored after 3.11?) Tokyo: Chikuma-shobo.
Tateishi, Y. (2014, September). *Houshasen hibaku mondai ni okeru kagaku kenkyu to hihan no ryoritsu—Kenkyu ryoiki goto no chigai ni tyumoku shite* (Compatibility of criticism and scientific research and criticism: Focusing on the differences in research area, Trans.). Paper presented at the 3rd Annual Meeting of the Japan Association for Science, Technology and Society.
Tsuchiya, Y. (2013). *Nihon ni okeru gan no hokandaitaiiryou no shimbun houdou ni kansuru kenkyu* (Analysis of newspaper coverage about complementary medicine for cancer in Japan, Trans.). Dissertation, Waseda University, Tokyo.

Wynne, B. (1991). Knowledge in context. *Science, Technology and Human Values, 16*, 111–121.
Wynne, B. (1996). Misunderstood misunderstandings: social identities and public uptake of science. In A. Irwin & B. Wynne (Eds.), *Misunderstanding science? The public reconstruction of science and technology* (pp. 19–46). Cambridge: Cambridge University Press.
Waltner-Toews, D. (2008). *Food, sex and salmonella: Why our food is making us sick*. Vancouver: Greystone Books.
Wien, C., & Elmelund-Præstekær, C. (2009). An anatomy of media hypes: Developing a model for the dynamics and structure of intense media coverage of single issues. *European Journal of Communication, 24*, 183–198.
Xiong, Q. (2014). *Nihon no Komemondai ni taisuru shimbun houdou to nationalism* (Nationalism behind the rice articles in Japanese newspaper, Trans.). Dissertation, Waseda University, Tokyo.

Chapter 4
Rhetorical Marginalization of Science and Democracy: Politics in Risk Discourse on Radioactive Risks in Japan

Hideyuki Hirakawa and Masashi Shirabe

Abstract This chapter analyses "politics in the risk discourse of radioactive risks" that we have witnessed since March 11, 2011 in various discursive arenas such as the mass media, governmental/municipal decision making and risk communication activities, and arguments by individual scientists on Social Network Services (SNSs). The discourse has rhetorically marginalized what has been at stake in terms of public anxiety and controversies over the risks of low dose radioactive contamination of foods, water, soil, and tsunami debris. Such marginalization can be classified into three forms in terms of how the risk discourse downplays the significance of scientific and/or social dimensions: (1) Reduction in dimensions of issues to scientific ones and the problem of public misunderstanding of science (scienceplanation); (2) Mobilization of shaky or imbalanced scientific arguments; and (3) Emotional mobilization. We present eight case studies to exemplify these three forms of rhetorical marginalization of science and democracy in the risk discourse. In any forms of marginalization, legitimate democratic deliberation as well as genuine scientific arguments have been suppressed and replaced by top-down technocratic decisions that have sometimes relied on shaky scientific bases. In conclusion, we discuss the nature of these problems from the perspective of risk governance of technological disasters and reflexive questions as to the grounds of our criticism of marginalization.

H. Hirakawa (✉)
Osaka University, Osaka, Japan
e-mail: hirakawa@cscd.osaka-u.ac.jp

M. Shirabe
Tokyo Institute of Technology, Tokyo, Japan
e-mail: shirabe.m.aa@m.titech.ac.jp

4.1 Introduction

Since the explosions at the Fukushima Dai-ichi (No. 1) Nuclear Power Plant (NPP) of the Tokyo Electric Power Company (TEPCO), caused by the Great East Japan Earthquake and subsequent huge tsunami, Japanese civil society has gone through many adverse experiences. Radioactive fallout from the reactors has contaminated broad areas of eastern Japan. Although it is unfortunate that thousands of people who lived in some towns and villages near the Fukushima plant cannot return home for several decades or more, it is fortunate that the contamination levels in most areas are relatively lower than what was expected and the health risks posed by contamination have been deemed to be moderately less as well.

However, such situations where radioactive risks are considerably smaller than those posed by the previous tragedy of Chernobyl could themselves be a source of social unrest. If the risks were high enough, science would be able to provide the Government and the public with definite cautionary advice on countering the risks. However, in cases where risks are not very high, science cannot do that; rather on the contrary, it is prone to play a *suppressive* role against the public, especially those who care about the risks and related problems (Fig. 4.1). In other words, the situation of "small" risks has created a discursive arena of risk governance in which politics as to how to frame and address the problem of risks has taken place.

This is what we have witnessed since March 11 in various discursive arenas such as the mass media, governmental/municipal decision making, risk communication, and arguments by individual scientists on Social Network Services (SNSs). That is, the "politics of risk discourse" has rhetorically marginalized what has really been at stake in public anxieties and controversies over the risks of low dose radioactive contamination of foods, water, soil, and tsunami debris. As will be argued in the Sect. 4.2, such marginalization can be classified into three forms in terms of how the risk discourse downplays the significance of scientific and/or social dimensions. Consequently, legitimate democratic deliberation as well as genuine scientific

Fig. 4.1 Antinuclear demonstration in front of the Diet building

arguments have been suppressed in any forms of marginalization and replaced by top-down technocratic decisions that have occasionally relied on shaky scientific bases.

The first aim of this chapter is to exemplify these forms of rhetorical marginalization of science and democracy by presenting seven case analyses of risk discourse after March 11 (4.3). The examples are drawn from the events that occurred in the first year after the disaster (from March 2011 to March 2012). In addition, we also examine the marginalization of science and democracy in the process of implementing a new law in the eighth case analysis, the so-called "Victims Support Act", in the following years (4.4). Finally, we discuss the nature of problems of marginalization within the perspective of risk governance of technological disasters and reflexive questions as to the grounds of our criticism of marginalization in the conclusion (4.5).

4.2 Analytic Framework: Three Forms of Marginalization of Science and Democracy

As was previously noted, the rhetorical marginalization of scientific and social dimensions of the problems of risks posed by low-dose radiation can be classified into the three forms summarized in Table 4.1.

The first, form 1, is the reduction of wider dimensions of issues including social problems to a scientific dimension and the matter of public misunderstanding of science. Due to this reduction, various social, ethical, and political considerations concerning trust, responsibility, and democratic legitimacy that should be considered in making risk management decisions have been dismissed from public discourse in the mass media, governmental/municipal practices of risk communication, and policy making. What is really at stake involves diverse societal issues such as public distrust and responsibilities of the Government, TEPCO, and scientific communities, and public rights of participation, personal self-determination, and values of fairness and equity. However, these societal considerations have been rhetorically marginalized People's concerns about these broad dimensions of the issues have been reframed, or *replaced*, by a matter of understanding scientific concepts of risk, such as the magnitude of annual mortality rates. If people still feel anxiety about risk issues, they are to blame because they are ignorant of science. Examples of this form of marginalization can be more or less found in all our cases below. This form of marginalization can be called "scienceplanation[1]".

---

[1] *Scienceplanation* is a term coined by Shirabe, based on the concept of "mansplanation" which means the way a man comments on or explains something to a woman in a condescending, overconfident, and often inaccurate or oversimplified manner. Similarly, scienceplanation is defined as the way a scientist or an authority figure comments on or explains something to lay persons in a condescending, overconfident, and often inaccurate or oversimplified manner. See the definition of mansplanation (retrieved August 20, 2014 from http://dictionary.reference.com/browse/mansplanation).

Table 4.1 Three forms of marginalization of science and democracy

	Social dimensions	Scientific dimensions
Form 1: (All cases) Reduction of dimensions of issues to scientific ones and the problem of public misunderstanding of science (*scienceplanation*)	– Dismissing societal considerations concerning trust, responsibility, and legitimacy – Bypassing democratic decision making as such	
Form 2: (cases 1, 2, and 3) Mobilization of shaky or imbalanced scientific arguments	– Dismissing societal considerations concerning trust, responsibility, and legitimacy – Bypassing democratic decision making as such	– Marginalizing well-balanced and reliable scientific arguments and views – Deepening public distrust of scientists and the Government
Form 3: (cases 6 and 7) Emotional mobilization	– Dismissing societal considerations concerning trust, responsibility, and legitimacy – Bypassing democratic decision making as such	– Marginalizing well-balanced and reliable scientific arguments and views – Bypassing thorough scientific measurements and disclosure

The second form of marginalization (form 2) is mobilization of shaky or unbalanced scientific arguments in conducting risk communication with the public. The claim that radiation doses below 100 mSv are safe (case 1), employment of the theory of radiation hormesis in the argument of risk of low-dose radiation (case 2), and unbalanced adoption of the evidence of low-dose radiation risks (case 3) are typical examples. As a result, well-balanced and reliable scientific arguments and views have been marginalized, deepening public distrust of scientists as well as the Government.

The third form (form 3) is marginalization by means of mobilizing moralistic language that sounds unproblematic; on the contrary, it seems morally appropriate. For example, the Government has been appealing to public moral empathy with the sufferers of disasters by calling for support for them by accepting food, water, and debris from contaminated areas. "Solidarity (*KIZUNA*)" is part of the language used in this *emotional mobilization*. The sacrifice of this mobilization is not only legitimate democratic decision making and social justice for the responsibilities of relevant parties to be sufficiently fulfilled, but also reliable and balanced scientific views on the risks of low-dose radiation as well as sincere, thorough scientific measurement of contamination and disclosure of the results. Cases 6 and 7 are examples of this type of marginalization.

It has basically appeared that most marginalizing discourses in these forms have been based on paternalistic doctrine that public anxiety and fear against the risk of low-dose radiation should be erased. In reality, however, they have very often ended up by dividing public opinions between those in favor of and against alleged scientifically correct claims of risks, and they have sometimes aroused the outrage of people who are concerned about risks. In addition, the discourse has had an effect that has exempted the Government and TEPCO from their responsibility of taking appropriate measures to redress and mitigate damages and risks by diverting their responsibilities to individual consumers and farmers who have suffered from contamination. This aspect is also a source of anger by the public, especially those who are concerned about social, ethical, and political dimensions of risk issues.

We present seven case analyses to exemplify these three forms of rhetorical marginalization of science and democracy in risk discourse in the first year after the March 11 in the Sect. 4.3, and we also argue another case in subsequent years in the Sect. 4.4.

4.3 Case Analyses of Rhetorical Marginalization of Science and Democracy

4.3.1 Case 1: Propaganda that Radiation Doses Below 100 mSv Are "Safe"

A radiation health risk management adviser to Fukushima prefecture held a lecture in the city of Fukushima on March 21, 2011. The adviser, Professor Shun'ichi Yamashita, claimed in response to a question during the Q&A session: "Scientifically, there is no harm to human health if the external radiation dose rate is under 100 micro Sv/h." While the value of 100 micro Sv/h is an obvious mistake and was corrected later to 10 micro Sv/h which is roughly 100 mSv/year, his claim caused the following effect. Not only Professor Yamashita but also many government and municipal officials, politicians, and other experts have repeatedly delivered similar information to the public. They explicitly or implicitly suggested that dose rates of less than 100 mSv/year or cumulative radiation exposures under 100 mSv are harmless. After a while, their claims almost converged to the opinion that cancer risks by cumulative exposure to radiation under 100 mSv had not been proved scientifically.

What does this mean? Brenner et al. (2003) clearly explained it as follows.

> It seems unlikely that we will be able to directly estimate risks at significantly lower doses than these because of the practical limits of epidemiology discussed above. Of course, the fact that risks cannot be directly estimated at doses below, say, 5 mSv, does not imply any conclusion as to whether risks actually exist at these lower doses. As we discuss below, at lower doses inferences with regard to risk need to be based on understanding underlying mechanisms.

Then, they as well as radiation authorities like the International Commission on Radiological Protection (ICRP) and the Committee on the Biological Effects of Ionizing Radiation (BEIR) claim that the Linear Non-threshold (LNT) model of dose-response relation is a scientifically acceptable model. It is scientifically plausible to assume, especially for the purpose of protection against radiation, that very low exposures to radiation raise cancer risks in proportion to levels of exposures.

Contrary to the claim that the risks have not been scientifically proved, radiation exposures under 100 mSv are not necessarily safe. Actually, cancer risk has been evaluated scientifically. In this regard, this claim is a typical example of a case where scientific argument is marginalized by shaky scientific views, which is the second form of rhetorical marginalization in Table 4.1. However, it is simultaneously also a case where the extra-scientific arguments are marginalized by appealing to scientific arguments even if they are actually scientifically unwarranted, which is form 1 in Table 4.1 (i.e., scienceplanation). It is theoretically impossible to conclude whether radiation exposures in Fukushima prefecture are "safe" or "dangerous", or whether the risks are acceptable or not, solely by referring to science. To do so, we have to take into consideration not only scientific factors but also extra-scientific factors including value judgments. Even though it seems that safety is judged by science alone, it is in fact the result of scientists' own value judgments. In other words, the marginalization of extra-scientific elements by appealing to scientific arguments, whether they are scientifically reliable or not, means depriving people faced with risk of opportunities of judging for themselves whether exposures are safe or not. This is one of the reasons this sort of marginalization is problematic.

4.3.2 Case 2: Emergency Lecture Meeting Held by Science Council of Japan: "To Fear Radiation Correctly"

4.3.2.1 Intrusive "Emergency Lecture"

The Science Council of Japan (SCJ) held an emergency lecture meeting on July 7, 2011 entitled to "To fear radiation correctly". Its main objective was as follows.

> After the earthquake disaster, a lot of information about radiation has been released, and many people in this country are vaguely anxious about the effects of radiation on health. This emergency lecture meeting is intended to present correct information as it stands, to dispel anxiety about radiation, and to improve radiation literacy through lectures and panel discussions by leading researchers. (Present authors' translation)

The meeting was obviously designed on the "deficit model" (e.g., Wynne 1991) in communicating science to the public, assuming that the lack of clear and accurate knowledge about radiation was the source of people's anxiety. This way of thinking is a major source of the marginalization of extra-scientific, social dimensions of risks, i.e., scienceplanation. In the case of the Fukushima nuclear disaster, a survey of parents' anxiety after the accident (Tateno and Yokoyama 2013) actually revealed that people's anxiety was derived from distrust of the Government and the uncertainty of scientific information about low-dose radiation, unidentified hot spots, and food contamination that had not been monitored carefully, rather than a lack of knowledge. With multiple answers allowed, 45.2 % of 1793 respondents selected "distrust of the outlook and actions by the Government" as the predominant reason for their feelings of anxiety, followed by "hot spots and unintentional intake of contaminated food" (40.8 %) and "uncertainty of low-dose radiation" (40.7 %). The proportion of respondents who selected "lack of knowledge regarding low-dose radiation" was 29.2 %. Thus, it is obvious that providing people with more knowledge cannot ease their anxiety.

Of course, it is true that to provide scientific knowledge about relevant risks sometimes dispels vague feelings of anxiety, but in relation to the risks of radiation, it is difficult to tell what "correct" knowledge is. In the first place, even if there were such knowledge, we could not necessarily expect that the knowledge could ease anxiety about radiation. In fact, the current consensus among international authorities like ICRP and the United Nations Scientific Committee on the Effects of Atomic Radiation (UNSCEAR) on the health effects of radiation is as follows.

1. Cumulative radiation over 100 mSv significantly raises cancer risks.
2. There are no datasets large enough to epidemiologically test whether there are cancer risks of cumulative radiation under 100 mSv or not.
3. It is scientifically plausible to adapt the linear non-threshold (LNT) model, which is induced from current scientific knowledge like mechanisms of cancer development, to evaluate cancer risks.

Since the LNT model suggests that there is no threshold to distinguish safe and dangerous levels of radiation, we cannot determine any acceptable level of radiation solely through science. Thus, "correct" knowledge cannot necessarily lead to dispelling anxiety about radiation.

Nevertheless, judging from the objectives of the meeting, the SCJ expected their audience to learn to fear radiation *correctly* and to dispel any anxieties by understanding *correct* scientific information. In other words, the SCJ exhibited its view that it was incorrect to have anxiety about radiation in Japan at that time.

However, did anxiety about radiation result from a misunderstanding of radiation? Even if radiation risks had been measured scientifically, would it be incorrect for people to feel anxiety based on their own value judgments and social contexts in which they had experienced risks?

Concerning these questions, UNSCEAR presented different opinions from those of SCJ, based on their report (UNSCEAR 2000, p. 513) on the Chernobyl nuclear plant accident as follows.

> 384. Many aspects of the Chernobyl accident have been suggested to cause psychological disorders, stress and anxiety in the population. The accident caused long-term changes in the lives of people living in the contaminated districts, since measures intended to limit radiation doses included resettlement, changes in food supplies and restrictions on the activities of individuals and families. These changes were accompanied by important economic, social and political changes in the affected countries, brought about by the disintegration of the former Soviet Union. These psychological reactions are not caused by

ionizing radiation but are probably wholly related to the social factors surrounding the accident.

385. The decisions of individuals and families to relocate were often highly complex and difficult. The people felt insecure, and their lack of trust in the scientific, medical and political authorities made them think they had lost control. Experts who tried to explain the risks and mollify people were perceived as denying the risk, thus reinforcing mistrust and anxiety.

In this way, the committee is cautious of attempting to dispel public anxiety by using scientific knowledge without considering social factors. In other words, it warns that the marginalization of social dimensions of risks is itself a source of people's insecurity. Contrary to this careful attitude by UNSCEAR, the SCJ and many other scientists and science journalists in Japan, have mistakenly tried to mollify people by scienceplanation. Consequently, not a few members of the public have perceived them as "denying the risk, thus reinforcing mistrust and anxiety".

4.3.2.2 "Science" in Emergency Lecture Meeting Held by SCJ

The main problem at SCJ's meeting was not limited to the marginalization of extra-scientific factors behind people's fears and anxieties by framing these emotions as a matter of the lack of scientific knowledge, based on the deficit model. The meeting was also problematic in the light of science, viz., the organizer seems to have taken little care in balancing the topics of lectures. As a result, marginalized public anxiety might have been driven further in a certain direction.

Table 4.2 lists four lecturers and their foci at the meeting. There seem to be two or three problematic topics chosen by the organizer. The most "impressive" one was a lecture by Professor Kiyonori Yamaoka. His lecture was about radiation hormesis, which means that lower radiation has good effects on the health of cells and/or organs. At the end of his talk, he suggested that hormesis effects on human health were about to be scientifically proved.

In reality, the existence of radiation hormesis is disputable even if it is looked at with a favorable eye. UNSCEAR (2010) evaluated the hormesis and other relevant hypotheses as follows.

32. The induction and development of cancer after radiation exposure is not simply a matter of the stepwise accumulation of mutations in the DNA of the relevant cells. There have been studies relating to the following hypotheses: (a) that adaptation of cells and tissues to

Table 4.2 Lecturers and their foci in emergency lecture meeting

Lecturers	Foci
Michiaki Kai	It is important to reduce radiation doses as much as possible
Kiyonori Yamaoka	Low radiation doses may improve human health
Yasuhito Sasaki	ICRP recommends taking into account damage and benefits from measures to protect society and individuals from radiation
Tokushi Shibata	The risk of radiation is smaller than that of smoking

low doses of radiation might cause them to become more resistant to cancer development (adaptive response); (b) that the effects of radiation on the immune systems, which recognize and destroy abnormal cells, could influence the likelihood of cancer development; and (c) that radiation can produce changes that create long-lasting and transmissible effects on the stability of cellular DNA (genomic instability) and/or trigger the transfer of signals from damaged cells to their undamaged neighbours (bystander effects); both genomic instability and bystander effects have been suggested as possible factors that modify radiation-associated cancer risk. These and other modulating factors, such as the induction of inflammatory reactions, could serve to increase or decrease the cancer risk due to radiation exposure.

Radiation hormesis is associated with hypotheses (a) and (b), which are less likely in the list. The most likely ones (i.e., bystander effects and genomic instability) are usually used to explain that low-dose radiation may raise cancer risks. Nevertheless, the SCJ only selected hypotheses intentionally or unintentionally that were less likely but meant positive health effects of radiation as topics at their meeting. By so doing, they attempted "to dispel anxiety about radiation, and to improve radiation literacy, through lectures and panel discussions by leading researchers". In this respect, the SCJ's meeting was also a case of marginalization in which well-balanced scientific views were threatened by shaky or less warranted science (i.e., form 2 in Table 4.1).

4.3.3 Case 3: 100 mSv for Governmental Working Group on Risk Management of Low-Dose Radiation Exposure

The third case of marginalization was found in a report produced by the Government's "Working Group on Risk Management of Low-dose Radiation Exposure" (referred to as "WG" after this), particularly in how the WG evaluated the risks of low-dose radiation.

The WG characterized cancer risks of low-dose radiation as follows (Working Group on Risk Management of Low-dose Radiation Exposure 2011: 5).

> Risk of cancer development from radiation at levels of 100 mSv or lower is considered so slight according to international consensus that risk is concealed by carcinogenic effects from other causes. At such low levels, clear increased risk of cancer development from radiation is difficult to prove. Scientific methods other than epidemiological studies also are being utilized to elucidate cancer risk, but at present, such methods have not yielded unequivocal risk information for humans.

In relation to the causes of this difficulty of proving the risks of low doses, UNSCEAR (2006) identified insufficient sample size (i.e., statistical power) and systemic errors in surveys as well as confounding biases with other factors. That is, by only referring to confounding biases with other factors, the WG succeeded in demonstrating that the risks were smaller than they could have been.

Furthermore, the WG claimed that an epidemiological survey in a high natural background radiation (HNBR) area in India indicated that excessive risks of cancers were not observed in the group with cumulative radiation doses over 500 mSv. These data came from a presentation by Dr. Kazuo Sakai, who was a member of the WG.

The data were introduced in a slide, which quoted two papers. The first (Neir et al. 2009) revealed that the size of the sample group was very small from an epidemiological viewpoint. Thus, although the original paper should be taken seriously, it was epidemiologically inadequate to only select data from that small group. The second paper (Hendry et al. 2009) concluded that:

> Results of studies conducted in HNBR areas sometimes have been used to discuss the validity of the LNT extrapolation of dose–effect relationships in the low dose and low dose rate range. Some authors have interpreted the absence of an observed excess of cancers as support for a protective effect of radiation. However, as discussed here, null findings are generally not informative. Given the limitations of the HNBR studies conducted up to now, it appears highly desirable to not over-interpret findings. This review leads to the unavoidable conclusion that, with one exception, any assertions on studies of detrimental or protective effects from HNBR exposure appear premature.

Scrutinizing the details of the report and materials presented in the WG, we can see that its members did not necessarily treat data and evidence in a "correct" way. That is, by mobilizing unbalanced "scientific" evidence, the report denied the mainstream opinion (i.e., LNT) and "underestimated" cancer risks of radiation doses of less than 100 mSv. This is again an example of the marginalization of science by shaky science, and consequently, the opportunities of considering extra-scientific aspects in risk analysis of radiation doses of less than 100 mSv were also dismissed.

4.3.4 Case 4: Anxiety Suppressed: Scienceplainers on Social Media

We have seen numerous claims trying to dispel public anxiety about radiation through scienceplanation not only by expert voices in rather formal, institutional settings in the previous three cases but also in informal, and often anonymous, discourses in mass media and SNS. To make matters worse, we have witnessed numerous cases on Twitter and other social media, where scienceplainers have made fun of other people who seemed to fear radiation "excessively" with "incorrect" knowledge. Such a phenomenon was also observed after the Chernobyl accident. In typical cases, the label of "radiophobia[2]" was used to describe "excessive" fears about radiation.

As was explained in case 2, however, UNSCEAR has been taking such fears much more seriously. It insists that "fears" by people living in contaminated areas

---

[2] In Japanese Internet slang, radiophobia is called "radio-brain" because the pronunciation of "radio-brain" in Japanese (housha-no) coincides with "radioactivity (housha-no)". The word appeared in March 2011, immediately after the accident.

should not be dismissed as peculiar responses ascribed to lack of scientific knowledge and that these fears have to be treated as being existing and real.

Contrary to these well-balanced understandings of fears about radiation, notwithstanding in a high-tech country in the 21st century, people with unshaped anxiety about radiations have been "attacked" so as to suppress their emotions, which could be very stressful for them. Such a suppressive and violent use of science, or sometimes disguise by science, is also an instance of the marginalization of extra-scientific dimensions of risks by science.

4.3.5 Case 5: Marginalization of Socio-Political Concerns by "Risk Comparisons"

The typical form of scienceplanation can also be found in the discourse of "risk comparison," which is a variation of the claim "to fear radiation correctly" in case 2.

Since immediately after the accident, we have repeatedly seen and heard scientists and government spokespersons comparing the magnitude of risk of low-dose radiation with that of other kinds of risks such as those from X-ray examinations, CT scanning, smoking cigarettes, and driving cars. They have tried to do this so that people can easily understand the magnitude of unfamiliar risk, i.e., probability, by comparing it with familiar risks.

However, it is well known by researchers and practitioners of risk communication that risk comparisons tend to cause severe public outrage and distrust of experts and the governments using them (Covello 1989; Slovic et al. 1990). Indeed, when the Prime Minister of Japan and His Cabinet published an essay entitled "Happiness of grandparents: Another face of radioactive materials" on its website on May 12, 2011, for example, the essay invited fierce public criticism on the Internet. The essay was written by a senior radiologist, Dr. Keigo Endo, who is President of the Kyoto College of Medical Science. He explained the benefits of the medical use of radiation by telling an anecdote about his patient who had overcome thyroid cancer (Endo 2011). Public criticism emerged on Twitter immediately after its publication. According to the record created by a web service called Ceron.jp, which automatically aggregates tweets that contain a certain URL, almost all 260 tweets aggregated in relation to the essay were negative responses, expressing anger, embarrassment, and dumbfoundedness.[3] In addition, responses to a questionnaire administered to local residents, who worked as public health nurses, school teachers, health advisors, and dieticians in the city of Koriyama in Fukushima prefecture revealed how the respondents answered a question regarding risk comparison (The University of Tokyo 2013). Asked "what do you think of the

---

[3] See the record on the Internet Archive Wayback Machine: http://web.archive.org/web/20140829035539/http://ceron.jp/url/www.kantei.go.jp/saigai/senmonka_g6.html.

explanation of risk of radiation exposure by comparing it with lifestyle-related risks like smoking?", 40.7 % of 210 respondents selected negative answers ("I feel disgusted" and "I feel a little disgusted"), even though 85.5 % chose positive answers ("it is helpful to understand the risk" and "it is slightly helpful to understand the risk") to a question whether that risk comparison is helpful for them to understand the risks or not.

Why does risk comparison cause such public outrage and distrust? There are mainly two reasons. The first is the problem with the persuasive intent of those who use risk comparison. They often aim at easing people's fears and anxieties, and sometimes at persuading them to accept the risk in question. It is quite common to hear in the discourse of risk comparisons that risks that are smaller than, or comparable to, the ones already accepted should be accepted, viz., "You have already accept these risks, so why don't you accept this one that is smaller than the others?" People often perceive such claims to be intentions to underestimate the severity of risks and feel they are being forced to accept unnecessary risks. In fact, many tweets referring to Dr. Endo's essay and responses to the questionnaire in the city of Koriyama expressed anger at persuasive intents of using risk comparisons.

Another reason for the failure of risk comparison is its marginalizing effect to reduce complicated dimensions of risk problems to a single measure of probability such as expected annual mortality rates. As is well known in psychological studies of risk perception, risk means something to people more than probability (Slovic 1987). Even focusing on quantitative aspects, the perception of risk, or (un)willingness to accept risk, comprises diverse measures such as expected annual probability of injury or disability (not death), spatial extent, concentration, persistence, recurrence, population at risk, delay, maximum expected fatalities, trans-generational effects, and expected environmental damage (Covello 1991). Furthermore, risk perception depends on qualitative factors such as the severity of consequences, catastrophic potential, reversibility, familiarity, scientific uncertainty, voluntariness, controllability, equity, clarity of benefits, and institutional trust. Furthermore, it is important to stress that some of these dimensions are not merely psychological but are social, ethical, and political (socio-normative). Equity, voluntariness, and controllability (i.e. self-determination), for example, are matters of social justice. Therefore, if people refuse to accept the risk of low-dose radiation and claim 1 mSv/year must be attained, it is not necessary that they do so only because of health concerns. They may reject the risk also because they do not want to be forced to accept it.

In contrast to such a complicated nature of risk perception, risk comparison tends to ignore all the dimensions but a single measure such as expected annual mortality rates or reductions in life expectancy. Consequently, it often causes outrage in people who are concerned with these ignored dimensions. In fact, since March 11, 2011 we have repeatedly witnessed on Twitter those who have expressed anger at comparing the risk of low-dose radiation with that of X-rays, CT scanning, or taking long international flights on airplanes. Many tweets against Dr. Endo's essay criticized it for its disregard of socio-normative dimensions of risk perception. They accused the essay of neglecting various differences of risks such as the difference between risks that were accompanied by benefits (e.g., cures for disease) and those without benefits, the difference between risks that one can voluntarily determine to take or not and risks that are imposed by accident, and the difference between artificially-generated risks and naturally-given risks. The same criticisms were also found in the comments of the respondents to the Koriyama survey.

4.3.6 Case 6: Damage by "Incorrect" Rumors ("FUHYOHIGAI")

Scienceplanation has been found in a way of framing the problems of social damage of the accident as well as a way of explaining and understanding risks of low-dose radiation.

"FUHYOHIGAI" is a Japanese term to describe situations where (mainly economic) damage is caused by incorrect rumors. For example, when BSE infected cows were found domestically in Japan in the autumn of 2001, sales not only of beef but also of other products including seafood such as scallops in Hokkaido prefecture severely dropped. Of course, there were no infectious connections between beef and scallops, but consumers avoided eating the latter only because they were products of the same area, Hokkaido, which was one of main birthplaces of BSE cows in several cases. As a result, Hokkaido's economy of agriculture and fisheries was severely damaged. This is an actual case of FUHYOHIGAI.

The situation after the nuclear accident, however, is more difficult to address. Some foods produced in eastern Japan including Fukushima are really contaminated by radioactive substances such as cesium 137. While it is a blessing in disguise that the contamination levels of most foods are lower than the international standards set by the Codex Alimentarius Commission of the Food and Agriculture Organization of the United Nations/the World Health Organization (FAO/WHO), this is simultaneously a source of problems. Why?

It is very difficult in the first place to define the criterion of what is "safe" and to share that definition widely throughout society. The best criterion in a strict sense that could be unanimously acceptable is "risk-free", but that is nothing more than words on paper. The best feasible criterion is that foods should be safe if their contamination levels are equal to or less than those before a nuclear accident. Conversely, if the contamination levels are much higher than acceptable standards, it would be unanimously judged as being sensible to reject eating these contaminated foods. In both cases, it would be much easier for many people to reach a consensus on sufficient safety or risk levels.

However, the situation at the time in Japan, especially in its eastern areas, was just in the middle of both ends of the contamination levels, between "safe enough" and "risky enough". While in most cases the levels were lower than regulatory standards, they were still more or less higher than ever before. Unanimous agreement on safety cannot be expected in such ambiguous situations. Judgments inevitably involve a range of socio-political and psychological dimensions and vary according to the preferences, values, and social/psychological contexts of people facing risks.

Furthermore, what is most important but makes situations more complicated is that it is a basic human right for people in democratic societies to choose what to eat and what not to eat, whatever its risk may be. On the one hand, we have individual freedom to choose to eat risky foods if they are allowed to be sold according to regulatory standards. However, we are not compelled to do so. Whether to eat risky foods or is not a matter of the right to choose or free will, and not a matter of obligation or responsibility.

What effects does abuse of the word FUHYOHIGAI bring about in society in such difficult and complicated situations? First, it should be pointed out that using this term has been inclined to obscure the liability of TEPCO (and the Government) while shifting it onto consumers. Claiming "FUHYOHIGAI" implies that economic damage to people in primary industries such as agriculture, forestry, and fisheries has been due to the behavior of consumers mistakenly avoiding negligibly small or non-existent risk by ingesting contaminated foods. However as was previously explained, the question of safety is not only a matter of science but also a matter of individuals' preferences, values, and basic human rights. Despite that, if they are obliged to eat foods that they do not conceive to be safe and do not want to eat, such a situation in itself is a violation of their individual rights to choose and live peacefully. Consumers who have to face contaminated foods and are forced to make choices are also victims of the accident, just as food suppliers who suffer from economic damages are. TEPCO and the Government are liable for both kinds of damage. However, the expression of FUHYOHIGAI tends to obscure such responsibility.

To claim "FUHYOHIGAI" is one form of rhetorical marginalization of extra-scientific concerns by appealing to science in this regard. The underlying philosophy is that safety is a matter of fact that can and should be determined by science. There is no room for considering extra-scientific concerns such as individual preferences, values, rights, or responsibilities of those who are liable for damages. For them, the problem is public misunderstanding of science about safety, i.e., if people understand the safety of contaminated foods in a scientifically correct manner, they should be willing to accept foods less contaminated than standards set by science. That is a world full of matters of fact but without involving matters of human values and justice.

However, arguments on principles of democratic societies do not necessarily lead to practical solutions in the problems with contaminated foods. There is indeed economic damage to producers, and we need any "reasonable" regulatory standards and systems to ensure those standards are satisfied. Extra-scientific as well as scientific factors should be taken into consideration to meet these ends. The question of what is reasonable entails various extra-scientific considerations, including consumer's empathies and solidarity with their fellow victims in food supply, as well as deliberation, consensus building, and mutual respect by the people involved. However, abusing the term FUHYOHIGAI precludes a priori such considerations and political practices necessary for democratically as well as scientifically legitimate ways of problem solving.

With regard to this, ICRP (2007) recommended:

> (84) For management of the radiological quality of foodstuffs in a country with a contaminated territory, relevant stakeholders (authorities, farmers' unions, food industry, food distribution, consumer non-governmental organisations, etc.) and representatives of the general population should be involved in deciding whether individual preferences of the consumers should outweigh the need to maintain agricultural production, rehabilitation of rural areas, and a decent living for the affected local community. A thorough debate at national level is necessary to achieve a certain degree of solidarity within the country.

Fortunately, the abuse of the term FUHYOHIGAI has so far prevailed largely within discourses in mass media and SNS. The Government's official definition of FUHYOHIGAI ("rumour-related damage") incorporates extra-scientific considerations to some extent, and does not reduce the problem to a matter of misunderstanding of science (DRCNDC 2011, 149). Of course, we need to carefully scrutinize to what extent this notion has been applied to actual cases of compensation.

> The expression "rumour-related" damage is interpreted to mean a variety of things depending on the person, and it is sometimes used in the sense of damage caused by state of anxiety in consumers or trading partners who avoid buying goods and services due to concerns about risks even though there is absolutely no risk from radioactive material, etc. However, when we are speaking of a nuclear accident like the accident, we should instead at the very least, be considering "rumour-related" damage to be due to a rejection by the market in order to avoid the risk of contamination due to radioactive substances which are not necessarily well-defined in scientific terms. It follows that if this kind of avoidance behaviour can be regarded as reasonable, the damage will qualify for compensation.

4.3.7 Case 7: Marginalization of Science and Democracy by the "Emotional Mobilization"

The previously discussed cases are examples of marginalization of extra-scientific elements by reducing the dimensions of problems to those that are scientific (i.e., scienceplanation), or marginalization of science by introducing shaky, less-warranted "science", which are the first and second forms of marginalization in Table 4.1. This last case presents an example of the third form: it may be called "emotional mobilization" through which people try to mobilize other people to do certain actions by appealing to the latter's emotions.

Emotional mobilization is usually seen and condemned as conduct of those who emphasize, or sometimes exaggerate beyond scientific justification, how risky a certain technology or substance is. In appealing to people's fears, they try to gather wider support for their political position to accuse industry or the government in regard to the use of that risky technology or substance. Such acts are often condemned as emotionalism and sensationalism, especially by their opponents.

However, similar behavior can be seen on the side of opponents. The Government has conducted emotional mobilization in countering the problem of radioactive contamination in Fukushima. Its effect was to marginalize not only legitimate concerns for democratic decision making, civic rights, and the responsibilities of the Government and TEPCO but also genuine scientific arguments about the problem.

For example, we have repeatedly heard and seen slogans of "TABE-TE OUEN (to support by eating)" and "KIZUNA (solidarity)" since immediately after the earthquake and nuclear accident occurred, which were partly claimed as "countermeasures" against FUHYOHIGAI. A month after the disaster occurred, the Ministry of Agriculture, Fisheries and Forestry (MAFF) started a campaign called "TABE-TE OUEN SHIYO! (Let's support by eating!)" from April 15, 2011. It also started broadcasting a TV advertisement featuring a famous male pop group called TOKIO from June 8, 2011. The purpose of this campaign was very clear. It was aimed at promoting the consumption of agricultural and marine products in eastern Japan including Fukushima prefecture that had suffered from the disasters. By doing so, MAFF intended to help revitalize the local economy and people's livelihoods that heavily depended on agriculture and fisheries. TABE-TE OUEN and KIZUNA were the slogans for mobilizing the empathy of the public toward suffers, and appealing to the moral sentiments of people.

The TABE-TE-OUEN campaign would have been a beautiful story that proved strong solidarity (KIZUNA) among Japanese public. They would have willingly supported it. However in reality, the campaign has aroused public suspicion and criticism since its very beginning.

Why did the campaign face such fierce criticism? In the first place, food products from Fukushima and other neighboring areas have been more or less contaminated by radioactive substances. It appeared to people, particularly those who were seriously concerned about food safety, such as mothers of small children, that the government forced them to accept unnecessary, excess risks of radiation exposure. Since no measurements of food contamination were conducted sufficiently in the early phases after the accident, it was inevitable for consumers to avoid foods from contaminated regions. They had not been provided with enough information about contamination levels.

It also simultaneously appeared that the Government and TEPCO had shifted responsibility to consumers. Just as in the case of FUHYOHIGAI, slogans of TABE-TE-OUEN and KIZUNA have functioned as rhetoric to obscure the liability of TEPCO and the Government to pay compensation for economic losses to food suppliers. In addition, the Government and municipalities of stricken areas should have conducted thorough and reliable measurements of the levels of contamination in foods and farmland and disclosed the results to the public from the very beginning of the crisis. This is the only way to recover public confidence in food safety and risk management measures. On the contrary, they did not do so. For example, Fukushima Governor declared the safety of rice produced there in October 2011 based on the result of government-organized inspection. In the next month, however, rice samples contained radioactive cesium that exceeded the governmental standard were found one after another in several spots in Fukushima. Consequently, the reliability of the measurements done by the Government and local authorities was severely questioned. Consumer's avoidance of Fukushima's products was partly due to the fault of the Government and Fukushima local government.

And what is worse, the discourse of TABE-TE-OUEN has a rhetorical effect in that it renders the act of risk avoidance by consumers to be somewhat immoral. That is, those who rejected foods produced in Fukushima were selfish and made light of KIZUNA. In fact, one can easily find tweets on Twitter that denounce people who were concerned about contamination as practicing discrimination against Fukushima inhabitants.

The rhetoric of TABE-TE-OUEN and KIZUNA played a role to divert responsibilities in this way, from the Government, municipalities, and TEPCO. It marginalized rational scientific arguments about contamination as well as legitimate public concerns and expectations as to whether the Government, municipalities, and TEPCO bear the responsibilities for managing risks, while blaming the risk-avert public for being immoral. By doing so, viz., by bypassing legitimate but painstaking steps in politics and science, people were mobilized into consuming products from Fukushima in the cause of helping to revitalize sufferers. In a sense, the emotional mobilization might be seen as a Japanese variation of what Naomi Klein called "shock doctrine" (Klein 2008).

4.4 Science and Democracy After First Year of Post-March 11, 2011

As was previously discussed, we witnessed various forms of rhetorical marginalization of legitimate socio-political concerns by the public in the first year after the disaster, as well as genuine scientific arguments in relation to low-dose radiation risks. As more than two years have passed since then, has the situation improved? No. We are still experiencing similar deficiencies in science and democracy.

4.4.1 Establishment of Victims Support Act

The most notable issue in this regard was the "retardation" and "emasculation" of implementing the "Act on Promotion of Support Measures for the Lives of Disaster Victims to Protect and Support Children and Other Residents Suffering Damage due to the Tokyo Electric Power Company's Nuclear Accident" (Government of Japan 2012, June: referred to as the "Victims Support Act" after this or "the Act" in short).[4] We can see how extra-scientific as well as scientific concerns have been marginalized in the process of its implementation.

---

[4] Tentative translation of the Act by the Japanese Government is available at: http://www.japaneselawtranslation.go.jp/law/detail/?id=2279&vm=04&re=01&new=1. (As of October 2, 2014).

The Act was proposed by a caucus Diet member with strong appeals and the collaboration of the victims and their supporters such as NGOs and lawyers, and was finally established on June 21, 2012. Modeled after its precedent, Ukraine's 1991 law "On the status and social protection of the citizens who suffered as a result of the Chernobyl catastrophe," the Victims Support Act is to ensure the right of victims to avoid exposure to radiation by enabling them to choose any options of whether to move (evacuate) from, keep residing in, or return to the areas that Article 8 (1) defines as "Areas under Major Support Measures (referred to as the 'Areas under MSMs' in short after this)". In order to do so, Article 2 (2) of the Act requires that the Government shall take appropriate measures to support victims' lives regardless of their choice so that they can freely make their choices. The Areas under MSMs are defined in Article 8 (1) as "areas where radiation doses are below the level that requires the national government to issue orders for evacuation but exceed a certain level." The "level that requires the national government to issue orders for evacuation" is 20 mSv/year in accordance with the ICRP recommendations. While the evacuees from these areas have been compensated and supported by TEPCO and the Government, those voluntarily evacuating from, living in, or returning to areas outside the Evacuation Areas have received quite limited relief. Without the Victims Support Act, it would have been very hard for many of the latter people to exercise their right to avoid radiation because they could not afford to evacuate their living areas or have them sufficiently decontaminated.

The most notable feature of the Victims Support Act in terms of science and democracy is its recognition that victims' decisions concerning whether to evacuate or not as well as the Government's decision on setting of standard[5] with regard to the Areas under MSMs are *not a matter of science but a matter of policy* (Kawasaki and Fukuda 2013). The main reason for this is that these decisions cannot be made by science alone because of scientific uncertainties in the knowledge of health effects by low-dose radiation. In fact, Article 1 adduces it as the rationale of the Act by claiming "risks of radiation from said radioactive materials to human health have yet to be elucidated scientifically."

Current international scientific consensus is that there is no scientific evidence to prove the increase in other health effects than thyroid cancer that was confirmed by studies on the impact of the Chernobyl accident. This is the official view of the Japanese Government and its scientific advisors, i.e., mainstream radiologists. Additionally, the Government and mainstream radiologists share another view that the cancer risk of low-dose radiation at levels of 100 mSv or lower is so small that it is concealed by other carcinogenic effects, occasionally alluding that low-dose radiation is safe enough, as if 100 mSv is the safety threshold (see the Case 3 in Sect. 4.3). The Victims Support Act, however, opposes these views. Regarding the former view, it seriously takes the possibility that there may be other types of diseases than thyroid cancer such as heart disease, cataracts, and various chronic

---

[5] The "standard with regard to the Areas under MSMs" is referred to as "a certain level" in the stipulation of the Article 8 (1) in the Victims Support Act.

diseases, which is suggested by not a few studies.[6] Regarding the latter view, the Act does not underestimate the risk of radiation below 100 mSv. based on the LNT model, which is another international consensus shared by ICRP and UNSCEAR.

Based on such a precautionary attitude toward scientific uncertainties, the Act ensures that victims can exercise the right of self-determination of choosing whether to evacuate or not, so as to avoid radiation (Nakate and Kawasaki 2012; Kawasaki and Fukuda 2013). Furthermore, the Act also ensures the right to participate in policy making as well as transparency to ensure this right of individual decision making. The most important stipulation of the Act in this respect is Article 5 (3). It requires the Government to take measures necessary to reflect the opinions of victims in formulating the "Basic Framework", which is a set of measures for concrete implementation of the Act including the setting of standard for the Areas under MSMs. This setting of standard is very crucial for the Act to effectively provide relief to victims because it determines who is covered by the Act. The Victims Support Act does not leave this decision to science but to policy making based on stakeholder participation. By the same token, Article 14 requires the Government to carry out similar procedures with regard to formulating other measures.

In this way, the Act definitely aims at addressing the problem of exposure to low-dose radiation not as a matter of science but as a matter of policy.[7] Its significance is broad and not exhausted by ensuring the sufferers governmental support. It is also important that it justifies people's actions to avoid radiation exposure by evacuation, sufficient decontamination, or other means in terms of the right of self-determination (Nakate and Kawasaki 2012). Voluntary evacuees have been often criticized for their lack of scientific understanding of the risk of low-dose radiation, sometimes insulted by a pejorative word "radiobrain" (radiophobia). Critics believe that voluntary evacuees would have realized that they did not need to evacuate had they understood it scientifically. Their claims for evacuation and support were condemned as being scientifically irrational. The same has applied to

---

[6] One of the most cited documents is the report by the Ministry of Ukraine of Emergencies, *Twenty-five Years after Chernobyl Accident: Safety for the Future*, which is based on investigations on local residents living in area contaminated with low-dose radiation for many years (Ministry of Ukraine of Emergencies 2011). Its core parts, Chap. 3 and this chapter, have been translated into Japanese by the Chernobyl Health Survey and Health Care for Victims—Japan Women's Network, Citizen's Science Initiative Japan, and 10 volunteer individuals including the authors of this chapter. In addition, the Japan Broadcasting Corporation (Nippon Hoso Kyokai: NHK) produced a television program based on this report and conducted its own investigations in September 2012.

[7] There are other rationales for understanding the issues as a matter of policy while the promoters of the Act have not been mentioned. The acceptability of risk generally entails various extra-scientific considerations including value judgments as was argued in case 1 in Sect. 4.3. Decisions on whether to accept risks or not when radioactive contamination is caused by nuclear accidents or whether to evacuate or not, involves numerous questions concerning family income, jobs, children's schools, relationships with parents, relatives and neighbors, the meaning of life as well as health problems in the family, which are not limited to radiation risk.

people who are living outside government-instructed evacuation areas but have claimed for more all-out decontamination of living circumstances. The Victims Support Act warrants their claims and actions by providing the rationale in terms of human rights instead of science.

4.4.2 Basic Framework of Victims Support Act Finally Formulated, But ...

In this way, the Victim Support Act could have been a powerful legal and political means to improve the deficiencies in science and democracy as well as the situation of victims as such. In reality, however, the real situation is far from what was expected. Although Article 5 of the Act stipulates that concrete measures for its implementation shall be given by the Basic Framework, its formulation took more than a year. Meanwhile, the Liberal Democratic Party (LDP) defeated the Democratic Party of Japan (DPJ) in the 46th general election in December 2012 and returned to power.

Three months after the change in power, on March 15, 2013, the Reconstruction Agency, an organization responsible for implementing the Act, and other relevant Ministries jointly publicized "The Policy Package for Supporting Nuclear Disaster Victims" (Reconstruction Agency et al. 2013). Its main document maintained that it was based on the purport of the Victims Support Act, but it should be stated that it was hard to find the spirit of the Act in the package. While it comprised nearly a hundred measures, most of them were those already implemented by relevant Ministries and agencies. Even new measures were aimed at promoting the return of evacuees, not at helping people who hoped to continue evacuation or to start a new life elsewhere. Diet members who drafted the Act immediately held a public hearing meeting with the officials of the Reconstruction Agency on March 19, 2014, and severely criticized the package, pointing out, "there is no measure for voluntary evacuees, it aims to force them to return to Fukushima" and "it distorts the purport of the Act" (Hino 2014, 40).

In May 2013, a new report was presented to the 23rd session of the United Nations Human Rights Council, which could have been of much help for the promoters of the Act. It was a report by the Special Rapporteur, Anand Grover, on the Rights of Everyone to the Enjoyment of the Highest Attainable Standard of Physical and Mental Health (Grover 2013). It was based on Grover's visit to Japan from 15 to 26 November 2012, during which he exchanged information and opinions with relevant Ministries, municipalities, other institutions and civil societies. The report made various recommendations to ensure the "Right to Health" of the victims, and pointed out various deficiencies found in countermeasures taken by the Japanese Government. The recommendations included those urging the Government to implement the Victims Support Act. The promoters of the Act expected the report to accelerate the process, but it could not make a difference. On August

22, 2013, 19 victims (16 families) filed a suit against the Government at the Tokyo District Court on August 22, 2013 as a last resort to accuse it of retarding the formulation of the Basic Framework contrary to the Victims Support Act and to raise public awareness of this problem. The SAve Fukushima Children Lawyers' Network (SAFLAN) is a network of younger lawyers who mainly supported voluntary evacuees and backed these plaintiffs.

After all, it was on August 30, 2013 when the Government made public the draft of the Basic Framework and started public comment period. The framework, entitled "Basic Policy on Promotion of Measures for Supporting Victims in Their Daily Lives" (continuously referred to as the "Basic Framework"), was finalized and approved in a Cabinet meeting on October 11, 2013 (Cabinet Decision 2013, October 11). That was 471 days after the Victims Support Act had been enacted.

One of the serious problems with this issue was that it was not a mere delay in administrative procedures but rather intentional *retardation*. According to an article in the Mainichi Newspaper on August 1, 2013, which was based on information from an anonymous government official, the Reconstruction Agency had agreed with other ministries and agencies at a meeting on March 7, 2014 that they would postpone making decisions on which ministry or agency would take responsibility for setting a standard for the Areas under MSMs until the election for the House of Councilors in July 2013 because they were afraid that it would be too controversial and have an influence on the election (Hino and Hakamada 2013, August 1).

4.4.3 Further Problems in Substance and Procedures in Creating Basic Framework

The problem was not limited to hindering the formulation of the Basic Framework. Further problems could be found in its substance and procedures. In terms of substance, the draft of the Basic Framework had serious flaws and provoked fierce criticisms not only from the promoters of the Victims Support Act, such as the Diet Members Caucus for the Victims Support Act, the Japan Federation of Bar Associations (JFBA) and the Citizen Conference to Promote Our Act, a coalition that consisted of 63 NGOs, but also from local governments outside Fukushima prefecture.

According to the criticisms (Friends of the Earth et al. 2013, August 30; Hino 2014; Fukuda and Kawasaki 2014; Yamagishi 2013), the most controversial point was the definition of the Areas under MSMs in the draft. As was previously mentioned, the Victims Support Act defines these areas as those where the radiation level is below that which requires the Government to issue evacuation orders but above a certain standard. In other words, which areas were to be covered by the Act depended on how to set this "certain standard." The issue at stake was supposed to be on which level of additional radiation dose the line should be drawn. However, the definition proposed by the Government in the draft of the Basic Framework did not refer to the standard. Instead, it defined the Areas under MSMs in terms of administrative districts, viz., it designated 33 municipalities in Naka Dori (middle area) and Hama Dori (coastal area) of Fukushima prefecture as the Areas under MSMs. The Minister for Reconstruction, Takumi Nemoto, explained that the reason for defining the areas in this way was because demarcation in terms of radiation levels would divide communities (Minister for Reconstruction 2013, August 30). Contrary to this definition by the Government, the promoters of the Act had been claiming that the standard for the Areas under MSMs should be an additional exposure dose of 1 mSv/year (Ito 2013; Nakate and Kawasaki 2012). If this claim were adopted, the Areas under MSMs would have been much wider than those in the Government's proposal. It would have included large parts of Fukushima prefecture, Miyagi prefecture, Chiba prefecture, Tochigi prefecture, Ibaragi prefecture, Gunma prefecture, small parts of Saitama prefecture and the Tokyo Metropolis. These areas except part of Fukushima were excluded from the Areas under MSMs. While the Government added a new category called "Quasi-Areas under MSMs", it was just a nominal label for the areas where existing measures had already been enacted. In other words, this category would add nothing to the status quo. More than a dozen municipalities outside Fukushima expressed their criticisms through public comments and written opinions, and demanding papers in the names of assemblies, Mayors, or Governors against the Government's proposal.

Even for the Areas under MSMs, the Government only proposed three measures out of 120 measures in the Basic Framework. In the first place, almost three fourths of these 120 measures were identical to those of the policy package presented on March 15, 2013, which had already been severely criticized, as was previously mentioned. As a whole, the measures in the Basic Framework were exclusively aimed at promoting the return of evacuees to their hometowns. Much of the aid for housing and job finding was designed for evacuees who hoped to return home, whereas the measures for people wanting to continue to live in or to newly move to areas outside Fukushima were rare or short-term. All the measures related to risk communication were aimed at easing public anxiety and persuading evacuees to return home by providing correct scientific information and knowledge about radiation.

In fact, the Government's policy concerning evacuation had been exclusively directed at promoting the return of all evacuees from the beginning. It was December 2013 when the Government changed that policy and decided to aid evacuees to move and live in new places for the reason that there were too many areas around Fukushima Daiichi NPP where the radiation dose was too high to live (Cabinet Decision 2013, December 20). However, voluntary evacuation remained beyond its scope even in this new policy. Furthermore, in February 2014, the Reconstruction Agency and 10 other ministries and agencies adopted a new policy package for evacuees, but it was also aimed at promoting their return home. In fact, it was called the "Policy package regarding radioactive risk communication aiming for evacuees returning to their homes" (Reconstruction Agency et al. 2014; Government of Japan 2014, September). It was a matter of course that the Basic Framework was exclusively return-oriented.

The promise of the Victims Support Act that it would ensure victims' rights to avoid radiation by providing necessary support, regardless of their choice as to whether to evacuate or not, was severely violated in this way.

In addition to flaws in the substance of the Act, the process through which the Basic Framework was formulated was also highly problematic. The Reconstruction Agency woefully neglected the democratic process required by the Act. As was previously mentioned, the Victims Support Act stipulates that the Government shall take measures necessary to hear the opinions of victims in formulating the Basic Framework. People promoting the Act, such as victims, supporters, law makers, lawyers, and scholars, had understood, and still understand, this stipulation that "measures necessary" include stakeholder participation. Notwithstanding, the agency had not furnished any opportunity for public consultation except for collecting public comments on the draft of Basic Framework prepared by the agency and convening two-hour explanatory meetings twice, the first in Fukushima (September 11, 2013) and the second in Tokyo (September 13, 2013). Even with respect to public comment, its duration was only two weeks (from August 30 to September 13) at the outset, and afterward extended for 10 days (closing on September 23) because of severe public criticisms.

Until the day of the final decision, diverse stakeholders, individual victims, NGOs, lawyers, and municipalities, sent their opinions to the Government as public comments, written opinions, and demanding papers. They called for serious modifications to the measures but also claimed that the Government should follow more open and transparent processes before finalizing the decision. The total number of submissions of public comments was 4,963. According to Kousuke Hino, a journalist belonging to the Mainichi Newspaper, roughly 2,700 comments criticized the definition of the Areas under MSMs, while there were only two comments that approved the draft without any modifications (Hino 2014, 93). In the end, however, the Cabinet approved the draft almost without any changes on October 11, 2013, except for minor rewording. Thus, the Government broke the democratic promise of the Victims Support Act to ensure stakeholder participation and transparency as an essential part of the promise of the right to avoid exposure to radiation.

4.4.4 Underlying Conception Regarding Science and Democracy in Basic Framework

What was the underlying thinking of the Government on this "democracy deficit"? This was clearly reflected in the speech by Takumi Nemoto, the Minister for Reconstruction, at a press conference on March 22, 2013 (Minister for Reconstruction 2013, March 22). Asked about how to formulate the Basic Framework for the Victims Support Act, he answered as follows:

> The Areas under MSMs are defined as areas where the radiation level is below 20 mSv and above a certain standard, and I believe that the standard must be determined through

collecting technical, scientific, objective knowledge inside and outside Japan. In this respect, we asked the Nuclear Regulation Authority to do that. By reference to its contents, we want to make a decision on the standard for the policy support areas. (Present authors' translation)

Worse still, he responded quite negatively to a question as to whether the Government would hold meetings of public hearings and the like to formulate the Basic Framework: he only repeated an ostensible phrase, "We will hear the opinions properly when formulating the Basic Framework."

Thus, it is obvious that the Government's attitude toward the Victims Support Act was fundamentally opposed to that of its promoters. The Government framed the problem as scientific, or technocratic, where the promoters tried to define it as a matter of democratic decision making and self-determination. This is a typical form of marginalization of reducing social dimensions to scientific ones (form 1, scienceplanation, in Table 4.1).

However, there was simultaneously a twist in what followed Minister Nemoto's speech. What actually happened was that the Government defined the Areas under MSMs not in terms of scientific examination of the standard but through political considerations as was previously mentioned. It only replaced the participatory democratic decision making with traditional one that was closed, opaque, top-down, and bureaucratic.[8] However, it should simultaneously be stated that a hidden scientific assumption accompanies this replacement. It is an assumption that radiation doses below 20 mSv/year is sufficiently low so that there will be no statistically detectable increase in cancer risk and that there will be no other health impacts than those caused by mental stress by prolonged life under evacuation or fear of radiation. In this regard, the Basic Framework is opposed to the Act's purport of a precautionary approach to scientific uncertainties and by doing so, it also denies people's rights to make their own decisions on whether to accept the risks.

4.5 Conclusion

We have seen a lot of cases where various forms of rhetorical marginalization of science and democracy have taken place in the first year after the accident (Sect. 4.3). We also have looked into how the precautionary and democratic nature of the Victims Support Act has been marginalized and emasculated in the years that followed. In relation to democracy, the spirit of the Act that highly esteemed stakeholder participation and self-determination has been marginalized by recourse

---

[8] In the first place, the setting of 20 mSv/year as the standard for lifting governmental evacuation orders is the result of a closed political judgment, even though it has usually been explained as scientific. According to an article in the Asahi Newspaper on May 25, 2013, the Ministers in charge of addressing the accident initially argued whether to set the standard at 5 mSv/year, but they finally abandoned this idea because it would increase the number of evacuees and amount of compensation (Sekine 2013, May 25).

to science (i.e., the 1st form of marginalization, scienceplanation) and finally replaced by traditional decision making behind closed doors. The deliberation and negotiation by stakeholders and policymakers have been substituted by risk communication aimed at easing the anxiety of the public and at prompting evacuees to return home by disseminating "correct" scientific knowledge of radiation risks. This is also a form of scienceplanation. In relation to science, serious concerns about uncertainties have been marginalized, and the certainty of currently available knowledge has been repeatedly emphasized instead. This, again in relation to democracy, has served as an excuse for bypassing democratic procedures. The process of implementation, or precisely, emasculation, of the Victims Support Act was also a process of marginalization of science and democracy.

In concluding this chapter, we will examine the question of what the risk governance of technological disasters should be like, especially in situations where considerable disagreements and conflicts prevail in society as to the validity of scientific facts and the legitimacy of decision making. To clarify this, we also ask ourselves reflexive questions. We have criticized various forms of marginalization of science and democracy in this chapter. However, on what grounds is this criticism valid? If no damage to health by exposure to low-dose radiation occurs in the future, are the acts of marginalization of science and democracy judged unproblematic and does our criticism lose its grounds? In other words, is it possible to problematize the lack of precautionary and participatory processes as they are, irrespective of the consequences of governance? How can we justify our criticism of marginalization and claim for more precautionary, democratic, inclusive governance of technological disasters?

The situation with the problems can be described by using the classification of risks in terms of the quality and nature of available knowledge and information on these. The International Risk Governance Council (IRGC) has classified risk problems into four classes of "simple", "complex", "uncertain", and "ambiguous" in its *White Paper* (Renn 2005, 29–31). The latter three are characterized by specific difficulties in dealing with risk problems while the first is defined as the absence of these difficulties. Complexity refers to the difficulty of identifying and quantifying causal links between a multitude of potential causal agents and specific observed effects. Uncertainty often results from incomplete or inadequate reduction of complexity in modeling cause-effect chains. Finally, whereas uncertainty means a lack of clarity over the scientific or technical basis for decision making, ambiguity originates in divergent or contested perspectives on the justification, severity, or wider meanings associated with a given threat. Ambiguity is further divided into *interpretative ambiguity* (different interpretations of an identical assessment result: e.g., "does this result means that it is an adverse or non-adverse effect?") and *normative ambiguity* (different concepts of what can be regarded as tolerable referring to ethics, quality of life parameters, distribution of risks and benefits).

The problem with the situation of marginalization of science and democracy can be characterized by using this classification as conflicts in how to frame the risk problem of low-dose radiation exposure between risk classes, especially between those that are simple and ambiguous. Specific risk management strategies, relevant actors, and their types of discourses are allocated for each risk class (ibid, 15, 51–52) in this classification. Simple risk problems can be addressed using a "routine-based" strategy that draws on traditional, bureaucratic decision-making instruments with limited actors, mainly the staff of competent authorities. In contrast, ambiguous problems are better addressed by "discourse-based" strategies that seek to create tolerance, mutual understanding, and eventual reconciliation of conflicting views and values through the most inclusive participation. In terms of types of discourses between actors, discourse on simple problems is an "instrumental discourse" whose focus is mainly on technical aspects of the problems. The main task of risk communication is to provide the public, as well as relevant actors responsible for managing risk, with technical information on the nature of risk and how to manage it. Conversely, ambiguous risk problems need to be addressed in the "participative discourse" in which competing arguments, beliefs, and values are openly discussed.

It is obvious that the Government and other actors practicing marginalization have framed the problem of radiation exposure as simple, whereas people concerned with risk and promoters of the Victims Support Act have framed them as ambiguous as well as complex and uncertain. In other words, marginalization is an act of reducing ambiguous problems to simpler issues.

What is important here, particularly in relation to our reflexive questions, is that these conflicting framings are not logically equivalent. That is, the very existence of disagreements on how to frame the problem means nothing but that the problem is ambiguous and justifies precautionary, democratic, inclusive governance. This should be applied even if the problem appears simple for one party and the source of disagreement is deemed to be the others' misunderstanding or ignorance of scientific knowledge about risk. That diagnosis may turn out to be false in the end. Similarly, if there is no dissent at the outset by stakeholders over the diagnosis that the problem is simple, disagreement and conflict may emerge afterwards. In any case, precautionary and inclusive approaches should be taken as a *due process* of risk governance to first and foremost build consensus on the classification of the risk problem at the initial stage. If all the stakeholders initially agree that it is a simple risk so that its management can be entrusted to Government staff and small groups of experts, subsequent processes should remain transparent and open to public scrutiny and renegotiation in the future. IRGC's *White Paper* recommends setting up a screening board that comprises members of the risk and concern assessment team, of risk managers, and key stakeholders to perform this task (ibid 52–3).

However, why can the existence of dissent justify taking precautionary and inclusive approaches in the first place? Particularly for the inclusiveness, the numerous literature on STS as well as political science have provided rationales such as "normative", "instrumental" and "substantive" ones (Fiorino 1989). In short, the normative rationale is that inclusive political participation is the right of democratic society and instrumental rationale refers to the advantages of participation that it can serve to facilitate decision making and implementation of its results by reducing conflicts and improving trust in government agencies. Substantive rationale means that wider participation can contribute to producing better decisions by including knowledge, information, ideas, questions, concerns, expectations, and alternatives from a variety of sources. The rationale for taking precautionary approaches can be considered a vital part of these rationales of inclusion: prudence in addressing uncertainties means that public concerns for negative implications of uncertainty are properly considered in making decisions (substantive rationale), this may lead to reducing conflicts and improving trust in the Government (instrumental rationale), and therefore it should be part of political rights to participation (normative rationale). In summary, not taking inclusive and precautionary approaches means to run the risk of diminishing all these advantages in governance.

In addition, it is crucially important to see that non-inclusive and non-precautionary governance can cause *actual harm* to the victims of disasters. A survey done by Tsujiuchi and the Sinsai Shien Network in Saitama (SSN) (Tsujiuchi 2014) revealed that victims of the Fukushima nuclear accident have been suffering from serious mental stress and nearly or more than 60 % of evacuees who responded to the questionnaire (745 in Fukushima, 499 in Tokyo and Saitama in 2013) are possibly tormented with posttraumatic stress disorder (PTSD). The main cause of this suffering was anxiety concerning life expenses, compensation and habitation, job losses, and weakening relationships with neighbors. In other words, this was because of the lack of appropriate social care and support for them so that they could recover their lives, not merely the shortage of mental care, claims Tsujiuchi. Furthermore, the study found that the policies of the Government and municipalities in Fukushima to return evacuees to hometowns have intensified mental stress especially of those who are at a loss as to whether to return or who are afraid that they cannot return. According to Tsujiuchi, the rate of PTSD was higher for man-made calamities than those for natural disasters and, worse still, the delay in providing social solutions tended to prolong PTSD. In this respect, it can be said that the very lack of precautionary and inclusive approaches as due processes, in which victims' concerns and expectations are fairly and properly incorporated into policy making, was the cause of their situations worsening. The Government, industry and scientists have dealt with victims quite dishonestly in the history of pollution such as Minamata disease, so that it is quite easy for the victims of the Fukushima disaster to fear that the Government will also abandon them. This might be a part of the real mental damage to victims.

Finally, how can we get out of this severe predicament? It is too hard to answer this question. The marginalization of science and democracy in risk and policy discourse is partly a sort of *thinking habit* deeply embedded in the technocratic political culture of policymakers and experts. Its practice may largely be conducted unconsciously. In this respect, our analysis and criticism could serve to bring the problems to light of reflection and to improve the situation. However, marginalization is probably not a cause but an effect of other causes. It may partly, or largely, be an effect of traditional democracy deficits still prevailing in the Government and municipalities, most of whose bureaucrats are generally unfamiliar with inclusive approaches. Also, it may be due to the inappropriate design of processes to deal with scientific uncertainties and build a scientific consensus, including the ways of selecting experts as policy advisors, which victims' concerns over a variety of health impacts caused by the accidents, not merely by radiation, have been neglected. Furthermore, the problem may be a matter of *political will* to narrow the scope of policies to urging evacuees to return and, by doing so, to minimize the expense to compensate victims and assist in their recovery.[9] These *real* political problems are far beyond the scope of our present analyses. They are matters of real politics including social movements and politics inside the production of knowledge, which are still far from emerging.

References

Brenner, D., et al. (2003). Cancer risks attributable to low doses of ionizing radiation: Assessing what we really know. *Proceedings of the National Academy of Science, 100*, 13761–13766.

Cabinet Decision. (2013, October 11). *Hisaisha seikatsu shien to shisaku no suishin ni kansuru kihonteki na houshin* (Basic policy on promotion of measures for supporting victims in their daily lives), Government of Japan. Retrieved August 28, 2014, from http://www.reconstruction.go.jp/topics/main-cat2/20140526155840.html.

Cabinet Decision. (2013, December 20). *Genshiryoku saigai karano Fukushima fukko no kasoku ni mukete* (The policy for accelerating Fukushima's reconstruction from the nuclear disaster), Government of Japan. Retrieved August 28, 2014, from http://www.meti.go.jp/earthquake/nuclear/pdf/131220_kakugi.pdf.

Covello, V. T. (1989). Issues and problems in using risk comparisons for communicating right-to-know information on chemical risks. *Environmental Science and Technology, 23*, 1444–1449.

Covello, V. T. (1991). Risk comparisons and risk communication: issues and problems in comparing health and environmental risks. In R. E. Kasperson, P. J. M. Stallen (Eds.), *Communicating Risks to the Public: International Perspectives*, Dordrecht: Kluwer Academic Publishers, pp. 79–124.

DRCNDC (Dispute Reconciliation Committee for Nuclear Damage Compensation). (2011, August 5). Interim guidelines on determination of the scope of nuclear damage resulting from the accident at the Tokyo electric power company Fukushima Daiichi and Daini nuclear power plants, OECD nuclear energy agency. *Japan's Compensation system for nuclear damage: As related to the TEPCO Fukushima Daiichi Nuclear Accident*, OECD Nuclear Energy Agency, pp. 123–161. Retrieved August 24, 2014, from http://www.oecd-nea.org/law/fukushima/7089-fukushima-compensation-system-pp.pdf.

Endo, K. (2011). *So-fubo no shiawase: housei busshitsu no mou hitotsu no Kao* (Happiness of grandparents: another face of radioactive materials). Nuclear Disaster Expert Group, the Prime Minister of Japan and His Cabinet. Retrieved August 12, 2014, from http://www.kantei.go.jp/saigai/senmonka_g6.html.

Fiorino, D. (1989). Environmental risk and democratic process: A critical review. *Columbia, Journal of Environmental Law, 14*, 501–547.

Friends of the Earth et al. (2013, August 30). *Hisaisha no koe naki mamano kihon houshin an wa tetsuzuki ihan. Kizon seisaku no yoseatume wa mou takusan* (A joint statement of 27 NGOs: The draft of basic policy without the voices of victims is violating the procedure. No more patchwork of existing measures). Retrieved August 25, 2014, from http://www.foejapan.org/energy/news/130829.html.

---

[9] Ohshima and Yokemoto (2014) estimates the total costs for compensation, decontamination, restoration from accident, decommission of reactors, and other governmental countermeasures at 110.819 billion USD (1 USD = 100 JPY).

Fukuda, K., & Kawasaki, K. (2014, January). *Fuminijirareru 'hibaku o sakeru kenri': 'Genpatsu jiko kodomo hisaisha shienho' kihon hoshin o tou* (Overridden 'right to avoid radiation exposure': questioning the 'basic poloicy of the nuclear accident child and victims support Act'), *SEKAI,* No. 852 (special edition), pp. 122–131.

Government of Japan. (2012, June). *Tokyo denryoku genshiryoku jiko ni yori hisai shita kodomo o hajime to suru ju_min to_no seikatsu o mamoru sasaeru tame no hisaisha no seikatsu shien to_ni kansuru sesaku no suishin ni kansuru ho_ritsu* (Act on promotion of support measures for the lives of disaster victims to protect and support children and other residents suffering damage due to the Tokyo Electric Power Company's nuclear accident), Act No. 48 of June 27, 2012. Retrieved October 2, 2014, from http://www.japaneselawtranslation.go.jp/law/detail/?id=2279&vm=04&re=01&new=1.

Government of Japan. (2014, September). Events and highlights on the progress related to recovery operations at Fukushima Daiichi Nuclear Power Station. Retrieved September 30, 2014, from http://www.iaea.org/newscenter/news/2014/infcirc_japan0914.pdf.

Grover, A. (2013). Report of the special rapporteur on the right of everyone to the enjoyment of the highest attainable standard of physical and mental health, Mission to Japan (15–26 November 2012), the 23rd session of the United Nations Human Rights Council. Retrieved August 25, 2014, from http://www.ohchr.org/Documents/HRBodies/HRCouncil/RegularSession/Session23/A-HRC-23-41-Add3_en.pdf.

Hendry, J., et al. (2009). Human exposure to high natural background radiation: What can it teach us about radiation risks? *Journal of Radiological Protection, 29,* A29–A42.

Hino, K. (2014). *Hisai-sha Shien Seisaku no Giman (The deception of victims support policy),* Iwanami Shoten

Hino, K., & Hakamada, T. (2013, August 1) *Fukushima daiichi genpatsu jico hisaisha shien sakiokuri mitsugi* (Fukushima Daiichi Nuclear Accident A secret conference for postponing support for the victims), *Mainichi Newspaper,* morning edition in Tokyo, August 1, 2013.

ICRP. (2007). *Publication 111.* Amsterdam: Elsevier.

Ito, K. (2013, September) *Koushu no hibaku o nenkan 1 mSv ika ni* (Let us make the radiation exposure of the public below 1 mSv), *SEKAI,* No. 847, pp. 189–197.

Kawasaki, K., & Fukuda, K. (2013, September) *'Hibaku o sakeru kenri' wa naze gutaika shinai no ka* (Why has 'the right to avoid radiation exposure' not taken shape?), *SEKAI,* No. 847, pp. 179–188.

Klein, N. (2008). *The shock doctrine: The rise of disaster capitalism.* New York: Picador.

Minister for Reconstruction. (2013, March 22). *Nemoto fukko daijin no kisha kaiken* (Press conference of minister for reconstruction Nemoto), Reconstruction Agency, March 22, 2013. Retrieved August 25, 2014, from http://www.reconstruction.go.jp/topics/25322_1.html.

Minister for Reconstruction. (2013, August 30). *Nemoto fukko daijin no kasha kaiken* (Press conference of Minister for Reconstruction Nemoto), Reconstruction Agency, August 30, 2013. Retrieved August 25, 2014, from http://www.reconstruction.go.jp/topics/13/08/20130830201606.html.

Ministry of Ukraine of Emergencies. (2011). Twenty-five years after Chernobyl accident: Safety for the future, Ministry of Ukraine of Emergencies and All-Ukrainian Scientific Research Institute for Civil Defense of population and territories from technogenic and natural emergencies (ME of Ukraine).

Nair, R. R. K., et al. (2009). Background radiation and cancer incidence in Kerala, India-Karunagappally cohort study. *Health Physics, 96,* 55–66.

Nakate, S., & Kawasaki, K. (2012) Nihon-ban Chernobyl ho no kanosei to 'Hinan suru Kenri' (The possibility of Japanese Chernobyl act and 'the right to evacuate')," *GENDAI SHISO, 40* (9), 154–166.

Ohshima, K., & Yokemoto, M. (2014). *Fukushima Genpatsu Jiko no kosuto o dare ga futan suru no ka: saikado no ugoki nc motode shinko suru sekinin no aimaika to toden kyusai* (who will shoulder the costs of the Fukushima nu-clear accident? Obscuring responsibility and rescuing TEPCO). *Kankyo to Kogai, 44*(1), 4–10.

Reconstruction Agency et al. (2013) Genshi-ryoku saigai ni yoru hisaisha shien pakkeiji (The policy package for supporting nuclear disaster victims). March 15, 2014. Retrieved August 28, 2014, from http://www.reconstruction.go.jp/topics/post_174.html.

Reconstruction Agency et al. (2014). Kikan ni muketa hoshasen risuku komyunik êshon ni kansuru shisaku pakkeiji (Policy package regarding radioactive risk communication aiming for evacuees returning to their homes). February 18, 2014.

Renn, O. (2005). White paper of risk governance: Towards an Integrative Approach, International Risk Governance Council (IRGC). Retrieved July 19, 2014, from http://www.irgc.org/IMG/pdf/IRGC_WP_No_1_Risk_Governance__reprinted_version_.pdf.

Sekine, S. (2013, May 25). Kikan kijun genkakuka miokuru minshu seiken ji genpatsu hinan zou o kenen (Shelved stricter standard for return home, at the time of the administration of democratic part of Japan, being afraid of the increase of evacuees from nuclear accident), *Asahi Newspaper*, morning edition in Tokyo, May 25, 2013.

Slovic, P. (1987). Perception of risk. *Science, 236*, 280–285.

Slovic, P., Kraus, N., & Covello, V. T. (1990). What should we know about making risk comparisons? *Risk Analysis, 10*, 389–391.

Tateno, S. and Yokoyama, M. H. (2013). Public anxiety, trust, and the role of mediators in communicating risk of exposure to low-dose radiation after the Fukushima Daiichi Nuclear Plant explosion. *Journal of Science Communication*, 12, Retrieved June 25, 2013, from http://jcom.sissa.it/archive/12/02/JCOM1202%282013%29A03.

Tsujiuchi, T. (2014, January). *Shinkokusa tsuzuku genpatsu hisaisha no seishinteki kutsu: kikan o meguru kuno to sutoresu* (Mental anguish of Sufferers of Nuclear Accident continues to be severe: Distress and Stress associated with Return Home), *SEKAI*, No. 852 (special edition), pp 103–114.

The University of Tokyo. (2013). *Genshiryoku to chiiki jumin no risuku komyunikeshon ni okeru jinbun-shakai-ikagaku ni yoru gakusaiteki kenkyu seika hokokusho* (Interdisciplinary Studies in Humanities, Social and Medical Sciences on Risk Communication of Nuclear Power with Local Residents). Funded by the Strategic Research Initiative in the Foundation of Nuclear Power, Ministry of Education, Culture, Sports, Science and Technology (FY2012-14), Principal Investigator: Keiichi Nakagawa (University of Tokyo).

UNSCEAR. (2000). Sources and effects of ionizing radiation, United Nations.

UNSCEAR. (2006). Effects of ionizing radiation, United Nations.

UNSCEAR. (2010). Report of the United Nations Scientific Committee on the Effects of Atomic Radiation 2010, United Nations.

Working Group on Risk Management of Low-dose Radiation Exposure. (2011). Report: Working group on risk management of low-dose radiation exposure (temporary translation), office of the deputy chief cabinet secretary, Government of Japan, 22 December 2011.

WHO (World Health Organization). (2013). Health risk assessment from the nuclear accident after the 2011 great east Japan earthquake and Tsunami, based on a preliminary dose estimation, World Health Organization.

Wynne, B. (1991). Knowledge in context. *Science, Technology and Human Values, 16*, 111–121.

Yamagishi, K. (2013). 'Hisaisha seikatsu shien to shisaku no suishin ni kansuru kihonteki na houshin' ni kansuru kaicho seimei (Presidential statement on the 'Basic Policy on Promotion of Measures for Supporting Victims in Their Daily Lives'), The Japan Federation of Bar Associations (JFBA), September 11, 2013. Retrieved September 22, 2014, from http://www.nichibenren.or.jp/activity/document/statement/year/2013/130911.html.

Chapter 5
Public Participation in Decision-Making on Energy Policy: The Case of the "National Discussion" After the Fukushima Accident

Naoyuki Mikami

Abstract The summer of 2012 was a significant period for Japan's nuclear and energy policy in that, in response to the Fukushima nuclear accident, the government decided to undertake, what it termed, a "National Discussion" on energy policy. This was the first time that Deliberative Polling had been introduced in Japan on a policy level. This DP included 285 randomly selected participants from across Japan, asking them to deliberate on three nuclear energy policy options with the assistance of a panel of eight specially selected experts. The participants were asked to fill out a questionnaire at three points during the process, and their support for the three options (zero, 15, and 20–25 % nuclear dependency scenarios) were recorded and later analyzed. What was particularly interesting was the participants' desire not merely for information but also to challenge the panelists' individual viewpoints and to question the very framework of the discussion. Significant changes in the level of support for the three scenarios were observed, with approximately 50 % of the participants eventually choosing a shift away from nuclear dependency. An examination of the government's response to the National Discussion, including the DP, demonstrated that the results had a significant impact on nuclear policy, with the government explicitly stating that nuclear dependency should be phased out by the 2030s. A subsequent change in government led to the abandonment of this policy decision, but did not totally undermine the value of Japan's first attempt at combining a public participatory process with actual national policymaking in the field of science and technology.

N. Mikami (✉)
Hokkaido University, Hokkaido, Japan
e-mail: mikami@high.hokudai.ac.jp

© Springer International Publishing Switzerland 2015
Y. Fujigaki (ed.), *Lessons From Fukushima*,
DOI 10.1007/978-3-319-15353-7_5

5.1 Introduction

June 29, 2012, became a significant date for Japan's nuclear and energy policy, when more than 150,000 people[1] demonstrated around the Prime Minister's office to protest against the restart of two reactors at the Oi Nuclear Power Station in Fukui Prefecture. After the triple meltdown in Fukushima, nuclear power stations that had subsequently been stopped for periodic inspection had not been restarted, and the Oi restart represented the first attempt to reboot nuclear power stations in the post-quake period. Despite the surge of growing protest, the government went ahead with the restart two days later, but the cries for the abandonment of nuclear power became louder and louder in the aftermath.[2]

This day also has a place in history because the government presented three policy options proposing different levels of nuclear energy dependency by 2030, including a zero-dependency (0 %) scenario, and called for, what it termed, a "National Discussion" on the matter. A few months after the Fukushima nuclear accident, the government stated that it would "[s]timulate national discussion overcoming the confrontation between the opposition to nuclear power generation and its promotion."[3] After nearly a year of consultation with various experts and stakeholders, the government finally came up with the policy choices on June 29. As a means of National Discussion, the government introduced Deliberative Polling (DP), in addition to conventional public comment processes and public hearings, to generate informed and deliberated opinion from the general public rather than from inner stakeholders or experts. This public participation process was historic in that it was virtually the first opportunity for the general public in Japan to become involved in a debate over nuclear power and energy choices.

At the time of the disaster, 54 nuclear power reactors were in operation in this country, and the government had, in its 2010 Basic Energy Plan, just decided to construct at least 14 new reactors by 2030 as nuclear energy was regarded as the "backbone" of an energy policy that could best meet the simultaneous needs for energy security, environmental protection (reduction of CO_2), and economic efficiency. In addition, nuclear policy in Japan had long been characterized by a top-down, authoritarian decision-making process. The utilization of nuclear power had been consistently promoted as national policy even after a series of accidents in the 1990s and 2000s had undermined public trust in nuclear energy. The political, economic, and societal contexts that had preserved such a static nuclear policy are

---

[1] The organizers announced that 150,000–180,000 people participated on that day while the Tokyo Metropolitan Police estimated the crowd at about 17,000 people. (The Asahi Shimbun, June 30, 2012.)

[2] Hasegawa (2014) provides an overview of Japanese civil society's reaction to the Fukushima accident, including the demonstrations around the Prime Minister's office (so-called "Kantei Demonstrations").

[3] "Interim Compilation of Discussion Points for the Formulation of 'Innovative Strategy for Energy and the Environment'" (July 29, 2011, The Energy and Environment Council).

worthy of inquiry in terms of political science, sociology, and STS, and previous studies have focused on this situation (Honda 2005; Yoshioka 2011).

Instead of a thorough examination of the historical background, this chapter is devoted to the process by which the National Discussion on energy choices was advanced in an unprecedented manner after the Fukushima accident, with a particular focus on the Deliberative Polling on Energy and Environmental Policy Options (hereafter referred to as the Energy DP or simply the DP).[4]

5.2 Background to the National Discussion

5.2.1 Japanese Nuclear Policy and Public Participation

Nuclear policy in Japan prior to the Fukushima accident had been characterized by its "dual organizational structure" (Yoshioka 2011), which consisted of a coalition between the electricity industry and the former Ministry of International Trade and Industry (MITI) on one hand, and the former Science and Technology Agency (STA) on the other. The industry-MITI coalition mainly presided over the commercial use of nuclear power, with the STA controlling all other matters, particularly research and development projects. Throughout its history, there had been no role for public participation in nuclear policy, with the governmental and industrial elites dominating decision-making and excluding critical experts and stakeholders from a number of advisory committees. Nuclear policy had gone completely unchallenged in the Diet as well.

In the 1990s, however, due to repeated nuclear incidents and scandals, including the sodium leak at the Monju fast-breeder reactor (1995) and the Tokai-mura JCO nuclear accident (1999), some signs of democratization were observed in the nuclear policy. One such move on the part of the government was the Round-Table Conference on Nuclear Power Policy, which the Atomic Energy Commission set up after the Monju accident in the spring of 1996, "to seek the views of all levels and

---

[4] Regarding the DP and the National Discussion on Energy and Environmental Policy Options, several studies have already been published in Japanese. Sone et al. (2013), Yagishita (2014), and Yanase (2013) are detailed reports of the entire process by the DP organizers themselves, who are also researchers in deliberative democracy. Kobayashi (2012) and Yagi (2013) are personal reviews by members of the Independent Review Committee of the DP, and they point out a number of achievements as well as problems of the DP and the National Discussion process. Sugawara (2013), admitting the significance of the Energy DP, examines the DP process and results from the viewpoint of public opinion research, and he points out several problems, particularly in regard to the representativeness of the participants. Onai (2014) examines the National Discussion as an example of introducing the idea of deliberative democracy to nuclear politics, and criticizes the government for having so quickly abandoned the outcomes of the National Discussion after the regime change in 2012.

sectors of society in Japan, and to incorporate their diverse opinions as part of future nuclear energy policy."[5] The Round-Table continued for about four years until 2000 in different forms, during which time 23 meetings were held. The conference invited participation from various stakeholders, including those researchers and activists who were critical of nuclear energy, and provided a stage for discussions between pro- and anti-nuclear panelists. However, the sponsor of the Round-Table, the Atomic Energy Commission, declined to give details on how the discussions and subsequent results would be reflected in nuclear policy, and the participants' proposals were mostly ignored, particularly those advocating a moratorium on nuclear power (Onai 2007, pp. 83–89).

Another major occurrence during this period was a spate of regional protests against the national nuclear policy. It was in this context that the residents of Maki-town in Niigata Prefecture initiated a local referendum on the construction of a nuclear power plant in 1996. People in Maki held the referendum to decide for themselves whether they should approve the construction of a nuclear power station by the Tohoku Electric Power Company. The residents had a number of chances to discuss and consider the use of nuclear energy before they cast their ballots; thus, the local referendum can be said to have provided the residents with a chance to deliberate as well as participate (Onai 2007, pp. 89–94).

As a result of the referendum and the following decisions made by the municipal government, the Tohoku Electric Power Company was finally forced to abandon the construction of the nuclear power plant. Although the referendum in Maki-town was an unprecedented event in the history of nuclear policy in Japan, it did no more than stop the construction of one nuclear power plant, and the organizational structure supporting the nuclear policy was preserved.

The government was unsympathetic toward fundamental reform of the nuclear policy even after the Democratic Party of Japan (DPJ)-led coalition took power in 2009. The DPJ coalition government emphasized a politician-led initiative as well as the breakdown of bureaucratic domination in various fields, but the DPJ was more bullish with regard to the promotion of nuclear development and utilization than the former LDP (Liberal Democratic Party) administrations had been. It has been argued that DPJ's aggressive promotion of nuclear energy can be accounted for by its reliance to some extent on the Federation of Electric Power Related Industry Workers' Unions, which has been an important vote-getting machine for a certain faction within the DPJ (Yoshioka 2011, p. 355). It was only after the Fukushima accident that the government really buckled down to nuclear policy reform and public participation.

---

[5] The website of the Atomic Energy Commission. http://www.aec.go.jp/jicst/NC/iinkai/entaku/index_e.htm. Accessed on 20 Aug 2014.

5.2.2 From the Fukushima Accident to the National Discussion

In June 2011, three months after the Great East Japan Earthquake, the then Naoto Kan-led administration[6] established the Energy and Environment Council from among his cabinet ministers. According to its mission statement, the Council was supposed "to work together across the Prime Minister's office and ministries to institute an 'Innovative Strategy for Energy and the Environment,' comprising short-, medium-, and long-term strategies, which can remedy the disproportionate and vulnerable current energy system and respond to requests for energy safety, security, efficiency, and environmental protection." Immediately after the Council's establishment, the Prime Minister announced his policy to reduce dependence on nuclear energy and withdraw it completely at some point in the future. On July 29, two weeks after this statement, the Energy and Environment Council declared, as a part of cabinet policy, "The government would lessen the dependence on nuclear energy" and "develop a national discussion on 'how to materialize scenarios for reducing nuclear energy.'" In this way, the search for a method of reducing Japan's dependence on nuclear energy through "National Discussion" became one of the government's basic principles.

To provide a springboard for the National Discussion, the government spent nearly one year in preparing policy choices that provided the public with different alternatives for the country's reduction in nuclear power dependence (Miyagi 2014, pp. 6–17). To start with, the cabinet set up an expert committee to verify the actual costs of different power generation methods and, after two months' intensive discussion, the committee concluded, in its final report published in December 2011, that (1) nuclear power was at least 1.5 times more expensive than it had been claimed when allowing for political and accident risks, and that (2) thermal power would remain competitive for the time being despite the rising cost of fossil fuels. The report also emphasized that (3) renewable energy, such as solar and wind, could be achieved at prices as low as those of conventional sources of electricity if the economic and policy conditions were right. In summary, the report revealed that nuclear energy was not significantly cheaper than other methods of power generation, and other sources, including renewable energy, were sufficiently competitive to become major components of the country's energy composition.

On the premise of these estimated costs, the government next began drafting a set of policy alternatives related to the degree to which the country should depend on nuclear and other sources of energy for electricity. The proposed alternatives

---

[6] Kan was forced to resign in September 2011 due to the strong disapprovals of his post-quake recovery measures, which were indicated not only by the opposition parties and the public but also by some factions in the ruling party. Yoshihiko Noda took over and stayed in office until December 2012, when the DPJ suffered a crushing defeat in the general election.

Table 5.1 Three scenarios for 2030 (Based on "Options for Energy and the Environment," issued by the Energy and Environment Council on June 29, 2012)

	Share of nuclear energy (%)[a]	Share of renewable energy (%)[a]	Share of fossil fuels (%)[a]	Electricity generated (trillion KWH)	Greenhouse gas emissions (vs. 1990) (%)
Zero scenario	0	35	65	Approx. 1	−23
15 scenario	15	30	55	Approx. 1	−23
20–25 scenario	20–25	30–25	50	Approx. 1	−25
Pre-quake figures (2010)	26	10	63	1.1	−0.3

[a] Shares represent the proportion of electric energy generation only

involved not only energy composition but also nuclear fuel recycling and global warming prevention. To cover this wide range of issues, details were discussed in three advisory panels in different ministries: the Advisory Committee for Natural Resources and Energy in the Ministry of Economy, Trade and Industry (METI), the Atomic Energy Commission in the Cabinet Office, and the Central Environment Council in the Ministry of the Environment.

Although the cabinet at first intended to compile reports from these advisory committees and finalize the draft policy options in the spring of 2012, discussion in each committee dragged on until early summer. On receiving the final reports from the three advisory committees, the Energy and Environment Council drew up and presented the policy alternatives in a paper titled "Options for Energy and the Environment" on June 29, 2012. The alternatives consisted of three options with different degrees of nuclear power dependency, 0, 15, and 20–25 % (hereafter referred to as the zero, 15, and 20–25 scenarios), to be implemented by 2030, as shown in Table 5.1.

At the same time, the government advanced its plan on how to generate public discussion on these choices. In addition to conventional methods, such as public hearings and public comment, the government introduced DP for the first time in its official decision-making with the aim of better understanding deliberated public opinion.[7]

---

[7] A group of researchers (including the author), mainly those specializing in STS and participatory practices in science and technology, issued an emergency statement on the same day (June 29), warning that the government's DP plan did not allow sufficient time for preparation and lacked the independent steering body, both of which are necessary for the fair and proper organization of a mini-public. The researchers also insisted that the government clearly indicate how it would treat the results from the DP. Although the lack of time was unavoidable, the recommendation can in part be seen to have resulted in the establishment of the DP Steering Committee as well as the disclosure of the review process to be implemented after the National Discussion.

5.2.3 The Introduction of Deliberative Polling (DP)

DP is a participatory and deliberative method developed by James Fishkin, a US political scientist, which has been used in a number of countries around the world (Fishkin 2009). At the time of its introduction by the DPJ coalition government in 2012, there had already been five applications of DP in Japan, all of which had been initiated by various groups of social scientists as experimental projects. This time, the government officially sponsored DP for the first time, but it entrusted a specially organized steering committee with its implementation. The steering committee was composed of a handful of researchers led by Professor Yasunori Sone from Keio University, who is one of the leading experts in the method. Like other researchers in this area, he had been waiting his chance to put DP into practice while government officials searched for novel methods with which to realize authentic "National Discussion" on energy choices. The interests of the researchers and the government, therefore, can be seen to have coincided in this instance.

Generally speaking, DP consists of the five steps shown in Fig. 5.1. First, the organizer conducts a conventional opinion poll, asking several thousand randomly sampled participants to answer a questionnaire on a particular topic. This first questionnaire is usually referred to as T1 (T stands for "time"). In the Energy DP, T1 was conducted by means of a random digit dial (RDD) telephone survey, and nearly 6,849 people responded to 24 questions on energy choices. At the same time, as a second step, the organizer explains to the respondents that a deliberative forum

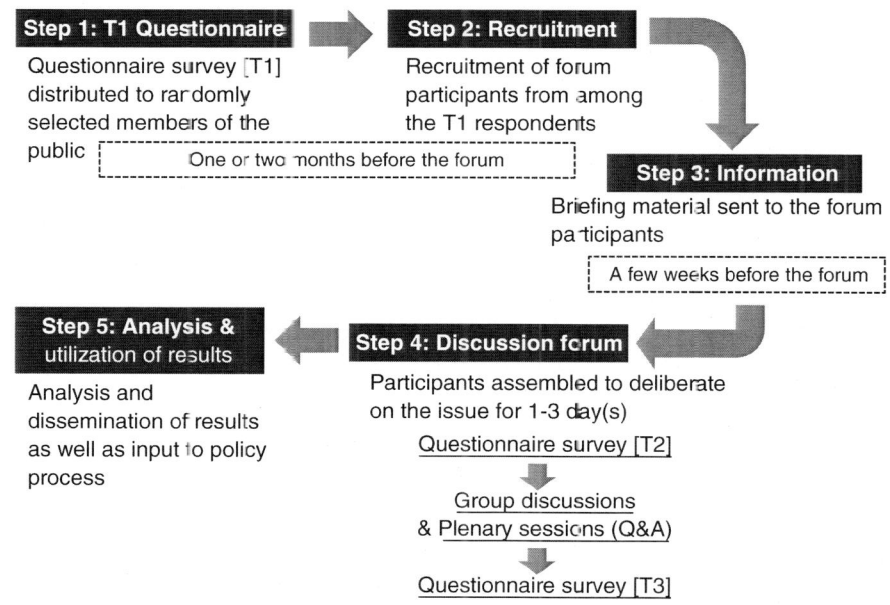

Fig. 5.1 DP procedure

on the same topic will follow and invites them to participate in the forum. In principle, the organizer pays the travel expenses of and honorariums to the participants. The idea is to gather various people (in terms of gender, age, region, ethnicity, education, occupation, religion, etc.) to create a microcosm of the society. It is desirable in most cases to recruit several hundred people, and in the Energy DP, 285 out of the 6,849 T1 respondents agreed to participate in the forum. Third, the participants receive briefing materials, usually a booklet, on the discussion topic from the organizer, and they are asked to read the materials to prepare for the forum discussion.[8]

Then, as the fourth step, the discussion forum is held. The forum lasts for one to three days. At the beginning of the forum, the participants again answer a questionnaire (referred to as T2) that includes the same set of questions as T1, and then the discussion starts. The participants take part in discussions in groups of fifteen, and they also have a chance to ask questions of experts on the topic in plenary sessions. Group discussions are facilitated by moderators, who are trained to withhold their own opinions under all circumstances and to help ensure fair and effective discussion. At the end of the forum, the participants are asked to answer the same questionnaire once more (T3), the responses to which can be regarded as more informed and deliberated public opinion than those of T1 or T2. Previous experience with DPs has shown that there is often considerable change in the participants' attitudes toward the issues after deliberation (Fishkin 2009, p. 134). Finally, as the fifth step, the results of the three polls are analyzed and published to provide reference material for policymakers and encourage further discussion by society as a whole.

5.2.4 Mini-Publics: Advantages and Disadvantages

It is obvious that the National Discussion on energy choices after the Fukushima accident was epoch-making when viewed in the light of the history described above. First of all, it is important to note that policy options included complete withdrawal from nuclear power. In addition, the National Discussion was designed so that the informed and deliberated voices of randomly selected citizens could be heard, through DP, as well as those of self-selected citizens and appointed stakeholders.

When considering this latter point, it is worthwhile to contrast two conventional methods for encouraging or visualizing public opinion; (1) self-selected methods, such as public hearings and public comment, on one hand, and (2) opinion polls of randomly selected members of the public on the other. Self-selected participants

---

[8] The English translation of the Energy DP briefing material (DP Steering Committee 2012a), along with other related resources, is available at the website of the Center for Deliberative Democracy, Stanford University. (http://cdd.stanford.edu/polls/japan. Accessed 20 Aug 2014.)

often have a lot of information and a chance to deliberate and are expected to provide clear-cut opinions that are worth listening to. However, their views are sometimes regarded to be extreme and not representative of the general public. Participants in opinion polls, on the other hand, can be seen as a representative cross-section of the entire population, but most respondents are chosen incidentally and are not ready to give an informed and deliberated opinion. The representativeness of the opinion and the deliberativeness of the participants are incompatible in most cases.

One of the breakthroughs in overcoming this dilemma is mini-publics (Goodin and Dryzek 2006), in which participants are gathered to form a microcosm of the entire society (a mini-public) by random sampling or some other ways, and asked to deliberate on various issues with balanced information at hand. DP is a typical method for the creation of mini-publics, and there are a number of participatory methods that fall into this category, as shown in Table 5.2.

Another well-known method of creating mini-publics is the consensus conference model, originally designed for participatory technology assessment (pTA) by the Danish Board of Technology in the 1980s. Consensus conferences have since become a big hit and have been used around the world with relation to a number of topics, particularly GMOs (genetically modified organisms) (Blok 2007; Dryzek et al. 2009; Einsiedel et al. 2001; Nielsen et al. 2007; Powell and Kleinman 2008). In Japan, STS researchers introduced this method in the 1990s and have organized conferences on a number of topics, such as GMOs, gene therapy, ICT, and nanotechnologies, as action research projects on public participation in science and technology (Kobayashi 2004; Wakamatsu 2010).

Table 5.2 Typical mini-public methods [Data from Shinohara (Ed.) (2012)]

	Length (days)	Number of participants	Origin (year)	Features
Deliberative Polling	1–3	Approx. 150–300	USA (1988)	Random-sampled participants fill in the same questionnaire before and after deliberation
Consensus Conference	3–8	Approx. 15	Denmark (1987)	A pTA method. Participants themselves draft a consensus document
Planungszelle	4	100 or more	Germany (1973)	Thorough deliberation, voting, and proposal drafting in each cell (five groups of 5 people)
Citizens Jury	5	Approx. 20	USA (1974)	Jurors issue a verdict after deliberation based on the testimonies of witnesses
Citizen Deliberation Meeting	1–4	Several dozen	Japan (2005)	Adapted and modified from the Planungszelle. Used at the grassroots level around Japan

Smith (2009) assesses various democratic innovations, including mini-publics, using four explicitly democratic goods; namely inclusiveness, popular control, considered judgment, and transparency (Smith 2009, pp. 72–110). Evaluated against these four criteria, mini-publics are thought to excel particularly in inclusiveness and considered judgment. By using random sampling and quota selection, mini-publics can involve a diverse group of citizens, and they also provide the participants with opportunities to receive a wide range of relevant information from experts and briefing materials. The participants also have the opportunity to understand different viewpoints and opinions as expressed by fellow participants.

In contrast, Smith argues that mini-publics have a number of weaknesses in terms of popular control and transparency. First, the agenda and corresponding experts invited to provide information are selected by the organizers or sponsors, not by the participants themselves, and the selection processes are out of popular control in most cases. Second, facilitators tend to have a strong influence on the discussion in mini-publics, and popular control is undermined if facilitators dominate important decisions, such as who should speak, which topic should be selected, when the discussion should end, and so on. Third, the political impact of mini-publics is quite limited and far from transparent. In most cases, the results of mini-publics are not linked to policy decisions, and it is difficult for participants of a mini-public, as well as the general public, to clearly predict how the results will impact relevant policy decisions. Fourth, the mini-publics generally receive little publicity, and most people, except the participants of the mini-publics themselves, are unaware of their actions or results.

5.2.5 The Analytical Perspectives of This Study

The abovementioned problems are common to the various applications of consensus conferencing in Japan, and, more generally, these can be regarded as fundamental challenges to the implementation of public participation and deliberation practices on science and technology issues. It can be said that they provide a useful perspective for the analysis and evaluation of the Energy DP.

However, the second point, facilitation, is rarely problematic as far as DPs are concerned as the quality control of moderators is always emphasized when implementing DP, and the moderators have to participate in training programs so that they can act impartially and dedicate themselves to the support of fair and effective discussions. The poor publicity of mini-publics, the fourth point, does not seem to apply in this specific case, either, for the Energy DP was conducted under an intense media spotlight (Ogiwara 2014).

On the other hand, the first and third points remain crucial in this particular case and represent a basic vantage point from which to observe and analyze it.

In regard to the Smith's first point, the expert-led agenda-setting was problematic in the case of the Energy DP as the three policy options for the National Discussion were prepared without any public consultation. The one-year process undertaken to prepare the options involved experts and stakeholders with different views on nuclear energy, and it appears reasonable in terms of process design that they should implement this process to narrow down the points at issue and draw up a limited number of viable options before opening up the discussion to the general public. The crucial point is whether there are any opportunities left in the public participation phase to challenge these preset agendas or policy options. Even if the agendas or options need to be provided as a starter, participants should not be limited to simply deliberating within the framework set by experts or stakeholders but should, instead, be allowed to question or challenge such a framework from within. To put it another way, the question here is whether the DP succeeded in providing substantial interaction between citizen participants and those experts who had hitherto played more significant roles in the decision-making process, and whether it was able to break away from token participation or a deficit-model in which experts dominate.

The other point, the impact of mini-publics on policy decision, has been among the most important issues in the implementation of such methods. This issue had been basically limited to the theoretical and hypothetical level, at least in Japan, particularly in regard to national policy decisions, as most mini-publics have been put into practice on an experimental basis or at a local level. In this sense, the DP on energy choices is exceptional in that it had a definite connection to the policy-making process. It, therefore, provides a valuable case study with which to examine how Japanese energy policy changed in reaction to the Fukushima accident as well as to assess the extent to which DP or mini-publics can influence policy decisions in the field of science and technology in more general terms.

The investigation of these two questions can also lead to an examination of several points at issue that have been the focus of STS research on public engagement to date. Previous research has demonstrated that dialogue in mini-publics represents the politics of public talk in which the 'public' is constructed, with a new form of deliberative and participatory governance contradicting the old one (Irwin 2006). As Felt and Fochler (2010) argued on the basis of empirical case studies of Austrian exercises, mini-publics frame certain roles and identities of the public at large while the participants sometimes try to resist and redefine such roles and identities. It has been observed that mini-public participants discover little connection to the wider political debate on the issues at hand and they tend to identify themselves as individual learners (Felt and Fochler 2010, pp. 234–235), and 'invited' participation, such as that of the Energy DP, tends to impose normative commitments on participants as to what is important and what is not (Wynne 2007). Further, in regard to the second point above, participatory practices provide mere promises of more democratic forms of science and technology governance instead of realizing a tangible effect on actual decision-making (Felt and Fochler 2010, pp. 235–236). How these existing challenges were met in the epoch-making attempt at an Energy DP are examined below.

5.3 Method

The Energy DP consisted of three questionnaire surveys, two group discussion sessions, and two plenary sessions (Q&A with invited experts). The results of the three questionnaires were tabulated, analyzed, and made fully available to the public by the DP organizer. The author used the data to gain an overview of the 'informed and deliberated' public opinion expressed through the DP.

At the same time, it is necessary to understand in more detail the nature of deliberation in the DP, particularly with regard to the research question of whether the DP provided participants with sufficient opportunities to challenge the preset framework and given policy options. To this end, it is desirable to have access to and analyze the records of the discussion. While the steering committee did not disclose the records of each group discussion, full video recordings of the two plenary sessions, approximately three hours in total, were made public through the official website[9] of the Energy DP. Based on these videos, the author transcribed the entire dialogue between the participants and experts, analyzed it qualitatively, and classified all 51 questions raised by the participants.

In addition, the author had a chance to observe the discussion forum on-site, being allowed to sit in on all the sessions including the group discussions.[10] The participants were divided into groups of fifteen so that 20 group discussions ran in parallel; thus, the author was compelled to select a few groups and listen to their discussions to confirm that they were being run appropriately. However, the author was able to gain full on-site access to the plenary sessions and later used field notes generated for analyzing the transcripts from the plenary sessions.

Apart from the analysis of the DP itself, it is necessary to answer how the DP results, together with those from the public comment process and public hearings, impacted the government's final decision. The government, on receiving the DP results, appointed an expert panel (the Review Panel on the National Discussion) for the specific purpose of reviewing the results of the National Discussion and deciding how best to apply them to the final decision with regard to the new energy strategy. The deliberations of this expert panel were completely public, with video recordings as well as handouts and minutes fully disclosed on the government's website.[11] This can be regarded as an exceptionally open and carefully thought-out example of the linking of public participation with the decision-making process. In view of the fact that the output from public participation in many other cases has been dealt with behind closed doors, the records of this Review Panel provide valuable data with which to consider the impact of mini-publics on policy

---

[9] http://www.cas.go.jp/jp/seisaku/npu/kokumingiron/dp/index.html. Accessed on 20 Aug 2014.

[10] The author worked for the Independent Review Panel of the DP, chaired by Professor Tadashi Kobayashi (Osaka University), which was commissioned by the Steering Committee to review the DP process. The author was appointed as an investigator for the panel and allowed to accompany its members to observe the entire DP proceedings.

[11] http://www.cas.go.jp/jp/seisaku/npu/policy09/archive12.html. Accessed on 20 Aug 2014.

decisions. From this perspective, the author scrutinized the minutes of the Review Panel and its final report, and examined how the DP results, as well as other output from the National Discussion, impacted the final decision.

In addition to these two major data sets, the author referred to various published documents, including those issued by the government and the DP Steering Committee, the memoirs of former government officials, and newspaper articles, in order to grasp a better overall picture of the National Discussion.

5.4 Details of the DP Discussion

5.4.1 Program Outline of the Energy DP

On Saturday and Sunday, August 4–5, 285 participants turned out at Keio University in Tokyo for the DP discussion forum. The participants had been randomly selected from the entire country and volunteered to travel to Tokyo to engage in the two-day discussion.

The discussion forum of the Energy DP began with an opening plenary session on the afternoon of Saturday, August 4. In the plenary session, the participants received guidance and responded to the questionnaire on energy choices for 2030 (T2 questionnaire survey). The participants were next divided into 20 groups of fifteen (from Group A to Group T) and were set to discuss the first topic, "Deliberations on Energy and its Judgmental Standard" for 90 min. At the end of the group discussion, each group selected a question that they wanted to set before the invited experts. Next, the participants assembled for a plenary session, in which the invited experts answered the questions from the 20 groups. After that, the conference dinner was held in the cafeteria.

On the next day, August 5, the exact same format of group discussion and plenary sessions was applied to the second topic, "Deliberation on the Scenario for Options in 2030." All the discussions finished at around noon, and before breaking up, the participants answered the final questionnaire (T3).

Before taking a closer look at the Q&A sessions, it would be helpful to devote some space to a discussion of the selection process of the experts. In general, the comments from experts at mini-publics have a great influence on citizen participants, and the organizers consider carefully who to invite as expert panelists. For example, in consensus conferences, participants have a say in the selection of both the expert panelists and the questions to be answered by them. Although the organizer select the experts in DPs, it is still necessary to achieve some balance in the composition of expert panels. On socially controversial issues, such as nuclear and energy policy, expert opinions are also diverse and are often diametrically opposed. It is a taboo to invite only experts sharing a specific opinion.

Further, it is important to balance the fields of expertise represented by the expert panelists. Most of the topics discussed in mini-publics are too complex to be dealt with completely by one expert, with energy and environmental policy being a typical example. Although covering all related areas is impossible, it is desirable to create an expert panel that can cover the major fields likely to be addressed by the questions from the participants. In the case of the Energy DP, it was necessary to invite not only experts on nuclear and renewable energy but also those with knowledge of the electricity systems, climate change prevention, the economy, and science and technology policy in general. However, there is usually very limited time for the Q&A sessions, and the discussion can become confused if there are too many experts. There are natural limits to how many experts can be accommodated in a mini-public, with four or five experts the maximum for a Q&A session (90–120 min in general).

Having met such conditions in creating an ideal list of prospective panelists, it is possible that some of them might decline the invitations due to scheduling or other problems. It is important to note that the experts who are present at each mini-public have been chosen in the face of a number of constraints. In the Energy DP, eight experts attended, four on Day One and another four on Day Two, and fielded questions from the participants.

Table 5.3 provides a list of the experts who attended the plenary sessions to answer the participants' questions. All of them had been deeply involved in the investigations into and establishment of countermeasures against the Fukushima

Table 5.3 Expert panelists invited to the Energy DP (Data from a DP discussion forum handout)

	Name	Title and affiliation[a]	Field of expertise
Day One (August 4)	OGIMOTO, Kazuhiko	Professor The University of Tokyo	Energy systems
	TAKAHASHI, Hiroshi	Research fellow Economic Research Center at Fujitsu Research Institute	Electricity and energy policy
	YAMAGUCHI, Akira	Professor Osaka University	Nuclear engineering; system safety engineering
	YOSHIOKA, Hitoshi	Executive vice president, professor Kyushu University	History of science and technology; science and technology policy
Day Two (August 5)	EDAHIRO, Junko	President Institute for Studies in Happiness, Economy and Society	Environment; communications
	SAKITA, Yuko	Journalist and environmental counselor	Sustainable community development
	TANAKA, Satoru	Professor The University of Tokyo	Nuclear engineering; nuclear waste management
	NISHIOKA, Shuzo	Research adviser Institute for Global Environmental Strategies	Environmental system analysis; global environmental policy

[a] The titles and affiliations are those at the time of the Energy DP

accident or had played a role in formulating the policy options for the National Discussion. As for the experts on Day One, Professor Yamaguchi of Osaka University specializes in nuclear power engineering and supports the continuation of nuclear power, while Professor Yoshioka of Kyushu University is a science historian who has studied the history of nuclear energy in Japan from an opposing perspective since the 1980s (Yoshioka 2011, p. 396) On the other hand, Professor Ogimoto of the University of Tokyo and Dr. Takahashi of the Fujitsu Research Institute were invited as experts distinguished for their knowledge of the energy and electricity systems as well as of renewable energy. In regard to nuclear energy, Prof. Ogimoto has argued that Japan should combine various electricity sources, including nuclear power, to realize economic efficiency and stability as well as safety, while Dr. Takahashi asserts that the reliance on nuclear energy has forced society to bear too great a burden in terms of both safety and the economy, arguing that it should be phased out in the future (Takahashi 2011, pp. 200–203).

The experts on Day Two all had key roles in the formulation of the three policy options, and they reflected the diverse opinions commonly held on nuclear energy. Ms. Edahiro, Ms. Sakita, and Professor Tanaka of the University of Tokyo were involved in the formulation of the three policy options as members of a branch committee in the Advisory Committee for Natural Resources and Energy, METI. Through committee discussions, it was obvious that Ms. Edahiro supports a zero nuclear energy option, whereas Ms. Sakita favors the 15 scenario, and Prof. Tanaka the 20–25 scenario. Dr. Nishioka of the Institute for Global Environmental Strategies was a member of the Global Environmental Committee of the Central Environment Council. He did not officially declare which option he favored, but he advocated the thorough utilization of renewable energy and energy conservation technology to realize a low-carbon society free of any dependence on nuclear energy (Nishioka 2011).

5.4.2 Q&A in the Plenary Sessions

Table 5.4 shows the participants' questions at the two plenary sessions classified by topic. Due to time constraints, each group was allowed to pose, in principle, only one question for each plenary session. In both plenary sessions, each group representative had a chance to ask a question that had been chosen at the end of the group discussion. The 20 groups were expected to ask one question each at the two plenary sessions, making a total of 40 questions. However, some of the questions actually consisted of two or three sub-questions,[12] and Table 5.4 counts all of them as separate questions, taking the total to 51 questions.

---

[12] For example, a question in the plenary session on Day One was presented as follows: "I'm sorry to trouble you with a similar question, but is there any prospect of the development of a renewable energy that is as efficient as nuclear energy? This is our question, and uh.... if we

Table 5.4 Classification of the questions raised by the DP participants[a]

		Technical questions	"In-depth" questions[b]		
			A	B	C
Nuclear power	Safety	2	3	–	–
	Cost and CO_2	2	2	–	–
	Back-end problems	2	2	–	–
	Restart of nuclear power stations	4	–	–	–
Renewable energy and energy conservation	Development and cost of renewable energy	9	–	–	–
	Energy conservation and CO_2 reduction	3	–	–	–
Electricity system and energy policy	Energy policy, nuclear energy and market economy	3	–	–	–
	Separation of generation and transmission	2	–	–	–
Scenarios	Zero nuclear option	2	–	3	1
	Process until 2030	4	–	–	–
	Other alternatives	–	–	–	7
Total		33	7	3	8

[a] Figures indicate the number of questions falling into each category
[b] The three types of "in-depth" questions are as follows
A: "Really?" questions about nuclear power
B: The feasibility of the zero scenario
C: The appropriateness of the three options

Q&A sessions in DPs and many other mini-publics are designed to provide participants with the opportunity to gain basic knowledge on the issues at hand. As shown in Table 5.4, two-thirds (33 questions) of the total questions asked the expert panelists for detailed technical information. In the Energy DP, the participants were provided with a 46-page briefing booklet that explained the background and details of the three policy options, and two-thirds (66.7 %) of the participants said they had read through it prior to the event. As is often the case with mini-publics, it was natural that the information from the booklet and subsequent discussion with fellow participants inspired the participants to ask questions aimed at obtaining further information about the subjects on the agenda.

At the same time, however, it is important to note that no less than one-third of the questions ("in-depth" questions in the table) went beyond such a basic role assignment in which lay participants ask for technical information and experts respond; in other words, the dynamic was not as simple as the experts teaching and

---

(Footnote 12 continued)
choose a zero nuclear option, the scenario says we need to increase the percentage of renewable energy from 10 to 30 %. What is the specific plan to realize this scenario?" (Group B).

the participants learning. The classification of the questions demonstrates that deeper interactions occurred on specific topics that formed the core of the discussion on national energy choice.

5.4.2.1 Day One: Questions About Nuclear Power

On Day One, the group discussion and subsequent plenary session were devoted to the topic of "Deliberations on Energy and its Judgmental Standard," which can be regarded as a prerequisite for discussing and selecting the best scenario for 2030. The corresponding chapter of the briefing booklet listed four perspectives as "fundamental criteria" (DP Steering Committee 2012a, pp. 14–18) for deliberation on energy and climate change issues; namely, safety, cost, stable supply, and prevention of global warming. The briefing material stated that these four criteria contradict each other, thereby presenting a "Quadrilemma" (DP Steering Committee 2012a, p. 19). The chapter then goes on to compare the strengths and weaknesses of nuclear energy, renewable energy, and fossil fuel energy, in light of the four criteria. Although it is desirable to phase out nuclear power from the viewpoint of safety, other sources have weaknesses in terms of cost and stability of supply (renewable energy) or prevention of global warming (fossil energy).

In the plenary session on Day One, nine out of the 20 groups asked in-depth questions regarding nuclear energy, ranging from requests for information on the safety and cost of nuclear power generation to inquiries on the disposal of high-level radioactive waste (HLW). Below are a few typical questions.

> We have heard that the Fukushima Daiichi Nuclear Power Station caused a severe accident due to the blackout caused by the tsunami, but did the earthquake motion really have no effect? (Group O)

> Are nuclear power stations in Japan really safe, especially in terms of technology and geographical conditions? Some of them are said to be on active faults ... (Group N)
> Does the amount cited in the briefing booklet really represent the true cost of nuclear power generation? Does the amount include the cost for final disposal as well as cooling, decommissioning, and repair? (Group A)

> I would like to ask whether nuclear energy really promises zero CO_2 emissions. (Group H)

It is interesting to notice that, as if by common consent, all the questions quoted above contain the word "really" ("*honto ni*" in Japanese). Another question in the same plenary was phrased as follows: "I wonder if HLW will be securely disposed of. While there is no prospect of restarting operations at Monju [a fast-breeder reactor], how do you think HLW can be treated?" (Group I) On a superficial level, this question appears to be asking for technical details about waste disposal methods, but it should be understood as a question asking whether HLW can "really" be disposed of securely as it was preceded by a remark expressing concerns about the safety of HLW disposal.

This wave of "Really (*honto ni*)?" questions should make us consider what the DP participants were actually seeking. As the above question from Group A ("Does the amount cited in the briefing booklet really represent the true cost of nuclear power generation?") clearly shows, basic information itself was provided, and most of the participants had read it in advance. These "Really?" questions were nevertheless asked as a kind of challenge to official explanations provided by the government, electric power companies, and academics.

Concerning the cost of nuclear power generation, the briefing booklet clearly states, in the figure on cost comparisons, that nuclear energy costs "9.0 yen or more per 1KWH" (DP Steering Committee 2012a, p. 13). This unit price was quoted from the report issued by the expert committee on cost verification (Sect. 5.2.2).

Until the Fukushima accident, the official estimate of nuclear energy cost was considerably lower at "5.9 yen per 1 KWH." However, the estimation did not include policy costs, such as the research and development budget or subsidies for accepting nuclear facilities, as well as social costs including accident-related expenses. The cabinet committee included these policy and social costs, and worked out the new unit cost of "9.0 yen or more."

Further, the estimation of electricity costs in the booklet was accompanied by a note stating, "The total amount of damage caused by the (Fukushima) nuclear accident has not been fixed yet, and the cost for accident risk stated here is a lower limit and can increase depending on the fixed amount of damage." Thus, only nuclear energy costs have no upper limit in the estimation.

Based on this, it is little wonder that the participants were skeptical about the integrity of the cost estimations. The experts seated on the stage were familiar with the discussions surrounding the electricity cost estimation, and the participants tried to assess how reliable the new estimation was by asking the question quoted above.

Each expert was asked to answer a question within about 2 min. This question on the "real" cost of nuclear power generation was answered by Prof. Yamaguchi, Prof. Yoshioka, and Dr. Takahashi.

The moderator called on Prof. Yamaguchi to answer first, and he emphasized that this was the best possible estimate at the moment.

> This estimate reflects various fluctuations including the uncertainty of cost factors. It has already included decommissioning or other related expenses mentioned earlier, and it can be regarded as the most accurate estimate that can be obtained at present.

On the other hand, Prof. Yoshioka highlighted the uncertainty surrounding the estimate. He started his answer by stating that he had "worked as a *goyo-gakusha* of the Atomic Energy Commission for thirteen years and was a member of the committees to formulate both the 2000 and 2005 Frameworks for Nuclear Energy Policy." *Goyo-gakusha* is Japanese slang usually used to refer scornfully to a scholar who is under the government's thumb, a lap-dog advisor so to speak. It is true that Prof. Yoshioka used to be a member of the government's committee on Frameworks for Nuclear Energy Policy, but he maintained a critical stance on the government's policy, particularly with regard to the nuclear fuel cycle program.

In this sense, the word *goyo-gakusha* here should be regarded as ironic, and Prof. Yoshioka answered the question from his experience as a *goyo-gakusha*.

> This should be understood as an estimate premised on a perfect situation in which every step of each scenario perfectly progresses without any accident or breakdown. This is a comparison [between various energy sources] in such a sense, so we should not trust it. (Laughs) The estimate in 2005 eventually turned out wrong because there has been little or no reprocessing [of spent nuclear fuel]. So it is always better to take into account risk when we think about the future.

Prof. Yoshioka spoke calmly and slowly, without raising his voice, and his words "we should not trust it (the government's estimate)" brought laughter for the first time in the plenary session and seemed to ease the tension in the auditorium.

Following these contrasting comments, Dr. Takahashi answered the question about the reliability of the estimated cost of nuclear energy.

> When we discuss economic efficiency, the single most important factor is uncertainty. The cost of nuclear energy was initially estimated at about 5.9 yen/KWH, but this was later changed to 8.9 yen/KWH,[13] according to the government's latest estimate. However, decontamination and decommissioning will continue for years to come, so it is quite uncertain how much the final amount will be.

Although Dr. Takahashi didn't explicitly state, "We should not trust it," he emphasized the uncertainty of the estimate just as Prof. Yoshioka had. His answer seemed to be somewhere in between those of Prof. Yamaguchi and Prof. Yoshioka.

Prof. Yamaguchi admitted that the estimate had a certain degree of uncertainty, whereas Prof. Yoshioka didn't say that the estimate was completely unreasonable. They agreed with each other that the estimate involved some kind of uncertainty. However, their comments appeared contradictory: "the most accurate estimate that can be obtained at present" (Prof. Yamaguchi) on one hand, and "we should not trust it" (Prof. Yoshioka) on the other. Thus, the single question whether the amount cited in the briefing booklet "really" represented the true cost of nuclear power generation revealed not only the information needed to make an informed judgment but also exposed the positions held by the key experts who had played important roles in generating the information.

5.4.2.2 Day Two: Questions About the Three Scenarios

The discussion topic on Day Two was "Deliberation on the Scenario for Options in 2030." The participants again began with group discussion sessions, and each group came up with a question. The plenary session then started, with four different experts invited, as described in Sect. 5.4.1.

---

[13] The cost verification committee estimated in the 2011 report that the cost of nuclear power was 8.9 yen/KWH or more. However, when the government recalculated this on the basis of the latest data for the presentation of the three options, it had increased by 0.1 yen to become 9.0 yen/KWH or more.

The participants again raised a number of in-depth questions that asked for something more than mere technical information, with such questions centering on two topics: (1) whether the zero nuclear energy scenario is possible, and (2) whether these three scenarios are appropriate from the beginning.

For the first topic, three groups raised questions about the possibility of implementing a zero nuclear energy scenario. For example:

> We would like to ask all the experts this question. Do you think the zero nuclear scenario is possible? And, what do you think Japanese society will be like when the zero scenario is realized? (Group A)

The question is clear and easy to understand, but it sounds a little strange when we consider the situation. The government presented the three policy options after a year of consultation with experts and stakeholders, and it was because the zero scenario was thought to be feasible and appropriate that it was included as one of the three options. The problem that must be now asked is why the participants questioned the feasibility of the zero scenario.

This demonstrates a structure similar to that observed on Day One in which some participants asked a number of "Really?" questions. The questions on Day Two should, in fact, be understood as follows: "There is a zero nuclear scenario in the briefing booklet, but we were wondering if it is *really* possible." However, there is a difference between this question and the "Really?" questions asked on Day One in that the zero scenario was not a focus of criticism or skepticism. A majority of the participants hoped that, in some way or another, the zero scenario would be a feasible solution, and wanted to ask the experts how realistic the zero scenario was. Here again, the real question was about the positions held by the experts.

If this is indeed the case, it might have been better for the participants to ask the experts directly which one of the options they thought the best. However, in DP and other mini-publics, it is basically the role of the participants to decide on policy alternatives, with the experts providing materials to assist their decision. Actually, the representative of another group (Group C) asked the same question, adding, "This might be a hard question to answer, but I would be happy to hear your answers," which shows that the participants were fully aware of this basic rule.[14] In summary, the question of whether the zero nuclear scenario is possible can be interpreted as a reflection of the participants' intention to maintain the initiative in the discussion as well as their desire to reveal the point of view of each of the experts.

The experts' answers revealed their individual ideas and principles regarding the utilization of nuclear energy. Ms. Edahiro clearly stated, "I think the zero scenario

---

[14] This observation is in agreement with the qualitative analysis of the group discussion records conducted by one of the DP organizers, Hironobu Uekihara. Based on his participant observation and analysis, Uekihara claimed that, in the group sessions on the morning of Day Two, more and more participants started to express their resolution to deliberate and decide on the energy choices for themselves, rather than depending on the experts and leaving decisions up to them. (Sone et al. 2013, pp. 162–181.)

is possible," and mentioned her experience on the government panel that helped formulate the policy options.

> The zero nuclear scenario is not something that is intended to attain zero nuclear power exactly in 2030, like a hole in one, but it means that we aim at the realization of zero, hopefully earlier than 2030, and even if it is delayed for some reason, for example, insufficient renewable energy, we should try for zero sometime as close as possible to 2030.

Ms. Edahiro highlighted the possibility that renewable energy will not be developed as fast as expected. However, she emphasized that the zero scenario can be attained if we use fossil fuel energy as interim relief, introducing as much natural gas as possible to avoid increases in CO_2 emissions, while decreasing nuclear energy dependence and strengthening renewable energy.

On the other hand, Ms. Sakita stated that the 15 scenario was realistic at that moment, while leaving open the possibility of a complete phase-out in the future.

> If we apply the forty-year decommission rule to the current nuclear power stations and close older reactors, the percentage of nuclear power will become about 15% in 2030. In the meantime, we should introduce as much renewable energy as possible to create vibrant life and communities. Also, it is necessary to promote energy conservation and reduce CO_2 from fossil fuels. After such efforts, sometime close to 2030, we should think seriously about how to deal with nuclear energy. If we think we can reduce nuclear energy then, we should change our direction.

Prof. Tanaka, a professor of nuclear engineering at the University of Tokyo, stated, "The zero nuclear scenario in 2030 does not seem to me appropriate as a national policy" due to the limitations of renewable energy and the soaring costs of oil and other imported resources. He also admitted, however, that a complete phase-out of nuclear power might be possible in 40 or 45 years.

Only Dr. Nishioka, an expert in global warming prevention, did not directly mention the feasibility of the zero scenario in 2030. Based on the discussion in the Global Environmental Committee of the Central Environment Council, he stated, "Technically, we can do without nuclear energy if we extend the scenario to 2050," adding that, in such a case, it would be necessary to make full use of the latest technology, such as CCS (carbon capture and storage), which would be a great challenge in itself.

The second point that resulted in in-depth questions on Day Two was whether these three scenarios were appropriate from the beginning. There were a total of seven groups raising this type of question, with one group asking a question belonging to this category on Day One. This means that at least one-third of the 20 groups wanted to challenge the given options.

These questions can be divided into two sub-categories: one in relation to why the options still included nuclear energy after the disaster in Fukushima, and the other in relation to whether 2030 was appropriate for such targets. The following question was typical of those in the first sub-category.

> We experienced severe damage due to the Great Earthquake on 3.11 last year, and nuclear energy doesn't seem necessary to us. Why do the options include nuclear energy? (Group P)

> If you promote deregulation, various potential renewable energy sources that have not been fully developed due to regulations can be utilized. And if technical and diplomatic problems with neighboring countries are solved, we can also use methane hydrate. If all these things are considered, there must be better options than these three scenarios. So I wonder why we have to choose from these three scenarios. (Group G)

These questions were answered together with other partially related questions and, unfortunately, the question as to why nuclear power still had its place in the policy options did not become the focus of the discussion. More precisely, the question from Group P was answered collectively with questions from two other groups on the safety of nuclear power. On the other hand, the question from Group G was treated together with a question from another group on electricity system reform.

Another focus of the questions about the three scenarios was whether 2030 was appropriate for the targets.

> Why are these options targeting 2030? We understand that the zero nuclear dependency in 2030 will have a great influence on us, and that makes us think it's impossible. Why aren't there any different options such as zero nuclear dependency in 2050? (Group S)

This question was actually treated together with the questions on whether zero nuclear dependency would be possible in 2050. As shown above, some of the experts commented that the zero scenario could be more realistic in the longer term; for example, in 2050. However, the question Group S asked was why 2050 was not the target despite such optimistic prospects, and this question was not finally answered.

There were a few other questions about the appropriateness of the target setting of the scenarios.

> If we think about the Fukushima accident, I think it possible to reach consensus on at least decreasing nuclear dependency. Don't you think it possible to choose such a policy as to reduce nuclear power by 1% each year? (Group Q)
> Why don't we have an option with a single-digit level [regarding the nuclear dependency in 2030]? We want a single-digit level option. Is it impossible? (Group K)

To the former question, Ms. Sakita commented that there wasn't any discussion about a scenario in which nuclear power would be reduced by 1 %, and there were no other comments on this question. To the latter question, on the other hand, Ms. Sakita and Ms. Edahiro reported that, during the scenario formulation process, some members argued that the dependency on nuclear energy might remain at a few percent in 2030 even if the zero scenario were chosen. Referring to the formulation process itself, they encouraged the participants to discuss whether the scenarios themselves were really appropriate, and not to regard them as something unchangeable.

> When we formulate the scenarios, we thought it necessary to narrow down the options for ease of discussion. Then, we finally came up with zero, 15, and more [20-25%]. If you think a single-digit option is essential, I would like to ask you to propose it. (Ms Sakita)
> The current discussion is moving toward a forced choice between zero and 15% in 2030. This makes me regret that we haven't kept a single-digit level option.... As Ms Sakita says, it is better to think that these scenarios are just a springboard for discussion and to voice your opinions by proposing alternative options or by declining the given options, instead of just choosing one from these three. (Ms Edahiro)

Here, the participants questioning of the scenarios and the experts responses to the questions finally appeared to communicate directly with each other. However, most questions regarding the appropriateness of the scenarios received insufficient responses. In general, the framework underlying the policy options was very obvious to the experts, and it was difficult for them to imagine how the participants would perceive it and what questions they would have. To deepen the discussion observed here, it is necessary to set up a different forum for discussion.

5.4.3 Participant Evaluation

At the end of the discussion forum, the DP organizer asked participants a few evaluation questions in the final questionnaire (T3). The participants were asked to rate each component of the forum, such as the plenary and group sessions, on a seven-point scale. To the question of whether "I learned a lot from people with different positions [in the group discussions]," 72.6 % of the participants answered positively (from 5 to 7), with 44.9 % answering "7 (strongly agree)." On the other hand, when asked whether "the responses of the experts [in the plenary sessions] were appropriate," only 21.4 % of the participants answered "7 (strongly agree)," although a total of 66.7 % of participants answered positively. The DP participants were quite satisfied with the dialogue between them and the experts while they also seemed to have gained inspiration from the group discussions.

In this regard, the results for the question of whether "the questions raised by other groups [in the plenary sessions] aroused my interest" also deserve some attention. About one-third (34.4 %) answered "strongly agree," with a total of 78.9 % of participants giving a positive evaluation. Both of these figures are more than 10 % points higher than those for question regarding the experts' responses. These results indicate that a variety of discussion points raised by other the groups were, at the very least, as informative as the responses given by the invited experts.

5.5 Impact on Policy Decision

5.5.1 Results of the DP and National Discussion

Figure 5.2 summarizes the participants' responses regarding the policy options (scenarios) in the three consecutive questionnaire surveys: T1, T2, and T3.[15]

---

[15] The questionnaires asked the participants whether they supported each of the three scenarios on a eleven-point scale, from 0 ("strongly disagree") to 10 ("strongly agree"), with 5 being "exactly in the middle." A "supporter" of a scenario here refers to a participant who answered exclusively in favor of one scenario, rating it at 6 or higher.

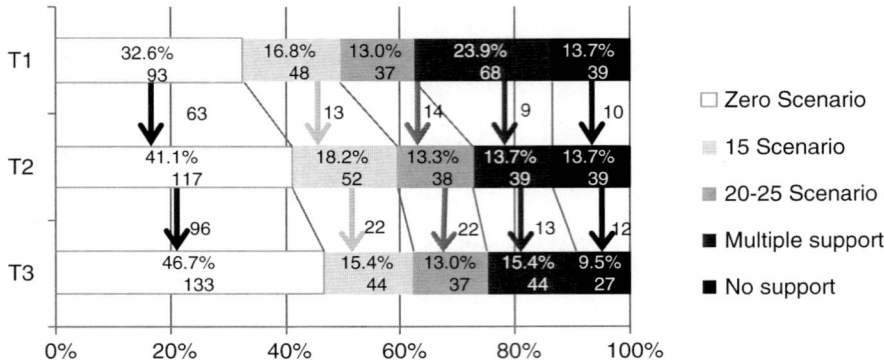

Fig. 5.2 Policy preferences of the DP participants (DP Steering Committee 2012b, p. 73)

The results show that more and more participants supported the zero scenario as they gained more information and had more opportunity for deliberation, with 46.7 % finally supporting the zero scenario. The percentages of participants supporting the 15 and 20–25 scenarios remained relatively unchanged, while the number of those who supported more than one scenario (multiple support) or who didn't support any one of the three (no support) decreased. On the surface, it appears that the participants' support shifted from multiple and no support to the zero scenario.

However, a closer look at the results on an individual level reveals that the changes in attitude were more complex. For example, only 13 (exactly a quarter) of the 52 respondents who supported the 15 scenario at T2 supported that option at T1. Similarly, only 14 of 38 participants (approximately one-third) who supported the 20–25 scenario at T2 had supported that option at T1. The attitude shift from T2 to T3 was less marked, but only half of the participants who supported the 15 scenario at T3 supported the same scenario at T2 (22 out of 44), while 22 out of 37 participants who supported the 20–25 scenario at T3 supported this scenario at T2. Thus, a significant percentage of participants, including those who eventually supported the 15 or 20–25 scenarios, demonstrated a change in attitude.

The questionnaires also contained items related to the criteria with which the participants decided on the appropriate policy option. The key questions in this regard asked participants to rank on an eleven-point scale (from 0 to 10) each of the four fundamental criteria discussed on Day One; namely, safety, cost, stable supply, and prevention of global warming. The T3 results for these questions demonstrate that safety was the participants' first priority, with 92.3 % answering they thought safety important (6 points or higher), with 78.2 % regarding stable supply, 60.4 % global warming prevention, and 48.4 % cost as important.

The problem that we have to consider is how we should interpret these results. In the DP report, which was published two weeks after the forum, the steering committee provided a discussion of the "policy implications" of the DP results (DP Steering Committee 2012b, p. 87), which affords a useful starting point for a consideration of the problem.

As to the reason for the exceptional emphasis on safety, the report states, "Public distrust, particularly of nuclear power, has not been dispelled yet," which appears to be an appropriate conclusion given the series of "Really?" questions asked at the plenary sessions. Support for the zero scenario steadily increased from T1 through T3, with the report claiming:

> The increase in the support for the zero scenario can be explained by the simplicity of the scenario. Choosing it compels us to pursue drastic growth in renewable energy and dependence on fossil fuel energy, which more and more participants seemed to accept as affordable after deliberation. (DP Steering Committee 2012b, p. 87)

The latter half of this quotation corresponds to our observations regarding the DP discussion. One of the most controversial parts of the discussion was whether the zero scenario was realistic. The overall DP results suggest that a considerable number of participants regarded the zero nuclear scenario as promising after two days of discussion. However, there is no conclusive proof that the participants found the zero scenario simple and easy to understand.

In fact, the analysis of the plenary discussions indicates that some participants thought the zero scenario was still ambiguous in a number of aspects, particularly in that it was not clearly explained why the target year was 2030 and how attainment of the scenario would affect Japanese society. As observed in the discussion on Day Two, the participants were not necessarily satisfied with the set of three scenarios given. Any one of the three scenarios was far from ideal for most participants. Under such circumstances, participants who voted for the zero scenario were forced to make a painful choice as that scenario was the only promising option for participants who regarded safety as the first priority.

When considering the impact of the DP on policy decision-making, we must also draw attention to the results of the other parts of the National Discussion (Fig. 5.3).

A total of 89,124 public comments were sent to the government, with an overwhelming majority (87 %) of the comments supporting the zero scenario. The government staff read all the comments and categorized them according to the major issue addressed in each comment. As a result of this analysis, the government reported that one-third of the public comments addressed three clear points; that is,

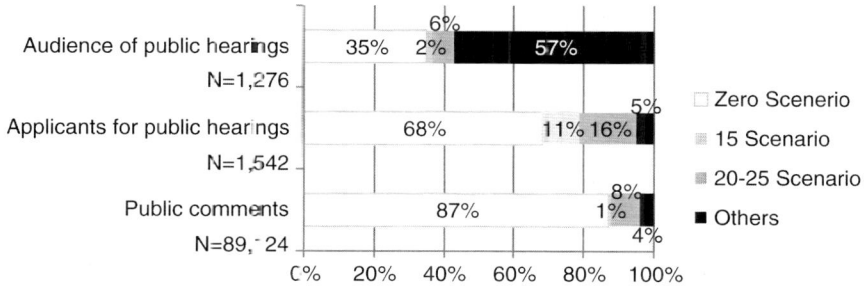

Fig. 5.3 Policy preferences expressed through public comments and public hearings (Based on materials distributed at the Review Panel on the National Discussion)

(1) concern about nuclear safety (47,901 comments), (2) the necessity for further development of renewable energy (35,063 comments), and (3) the unethical nature of nuclear development (33,276 comments). The government also reported the distribution of opinions expressed in the applications for and audience questionnaires of the public hearings held at eleven sites across the country, including Fukushima. According to the questionnaire distributed at the public hearings, the majority (57 %) of the audience did not clearly support a specific scenario. However, about two-thirds of those who volunteered as panelists in the public hearings supported the zero scenario. Although the preferences expressed through these self-appointed elements of the participation process do not necessarily represent the entire picture with regard to public opinion, it is important to note that a large number of people made an effort to directly express their concern about nuclear safety.

5.5.2 The Review Process of the National Discussion

At the time the government called for the National Discussion, it didn't specify how the results would be taken into account in the final policy decision. Generally speaking, it is almost impossible for policymakers to say with any certainty what impact a specific public participation process may have on decision-making, for public participation is only one of a number of elements influencing their decision. However, it is possible and even necessary to clarify at least the initial steps in the handling of the participation results; for example, detailing the advisory committees to which the results will be sent for further deliberation or the decision-makers who will refer to the results directly when formulating a final decision.

Although no prior arrangements had been made with regard to the process after the National Discussion, the government hurriedly set up an ad hoc expert panel, the Review Panel on the National Discussion, to deliberate on how best to interpret and use the results from the DP as well as those from the public comment process and public hearings. The Panel was chaired by the State Minister of National Strategy, Motohisa Furukawa, and consisted of eight experts in public opinion research, journalism, communication and media studies, political science, and administrative law. The Panel met three times between August 22 and 28 to discuss how to comprehensively interpret the results from the National Discussion as well as from other opinion polls conducted by the media, and how the government should reflect the results in the final decision. The meetings were broadcast live over the Internet, with all materials under discussion disclosed at the same time.

In the Review Panel, the secretariat (the National Policy Unit at the Cabinet Secretariat) first provided an overview of the results from the National Discussion, covering five sources: (1) DP, (2) public hearings, and (3) public comment, sponsored by the government, as well as (4) questionnaire surveys collected at fifty-four local briefings hosted by various private and civil society organizations, and (5) public opinion polls conducted by the media. The secretariat then proposed that these different results should be analyzed both quantitatively and qualitatively, not only counting or comparing the numbers supporting each policy option but also analyzing the logic behind the policy attitudes by scrutinizing answers to the DP questionnaires as well as comments and opinions expressed in the public comment process, public hearings, and questionnaires at local briefings.

The Panel members basically approved the proposal, and agreed that an emphasis should be placed on qualitative analysis in order to better understand the reasons for or logic behind the public's choices as well as the choices themselves. In reference to such complementarity between quantitative and qualitative analysis, there was active discussion about how to evaluate the public comments. Some members argued that numerical analysis of public comments was pointless as they did not represent public opinion at large. However, others emphasized the significance of the public comment process as a valuable means of public participation, arguing that the number of people supporting each policy option was to some degree meaningful.

In the second meeting, held five days later, the focus of the discussion moved on to how to interpret the results from the different processes, including the DP, as well as how to categorize major issues based on the arguments expressed in public comments or questionnaires from public hearing sites. One of the major issues here was the degree of reliability of the DP results. Some members challenged the legitimacy of the DP process, emphasizing its vulnerability to domination or manipulation; however, Professor Sone, who was the head of the DP Steering Committee and a member of the Review Panel, rejected such arguments. Although the panel members did not reach a clear consensus, most of them seemed to agree that, as one Panel member stated, "The DP can provide supplemental information about the thought of average people who do not actively participate in public hearings or public comment processes".[16] Another Panel member pointed out that the results clearly showed "people's outrage and distrust",[17] which also provided an important perspective when interpreting the output from the National Discussion.

The final meeting was held on the following day, and the secretariat presented a draft of the review report, which was approved after an hour and a half of discussion. The report summarized the results of the National Discussion and considered the implications for the decision of the energy strategy. Next, on September 4, the State Minister of National Strategy published its final version, which drew four implications from the results of the overall National Discussion.

First, the report concluded that more than a half of the population desires a move away from dependency on nuclear energy at some time in the future. Based on a detailed analysis of the DP results, almost half of those supporting the 15 scenario at T3 (about 7.4 % of the total participants) rejected the future utilization of nuclear power. The Minister, therefore, reached the above conclusion by adding this 7.4 % to the percentage of participants who supported the zero scenario at T3 (46.7 %) to give a slim majority of 54.1 %.

---

[16] Comment by Professor Junko Obata of Sophia University, a Panel member.

[17] Comment by Professor Tadashi Kobayashi of Osaka University, a Panel member.

Second, the report suggested that, despite this small majority, there was still debate over how fast Japan should phase out nuclear energy and how realistic the zero nuclear policy was. This conclusion was based on the fact that approximately half of the DP participants and about 50–70 % of the respondents to the opinion polls conducted by the media chose options other than the zero scenario, and that about half of the population expressed concern regarding the target of zero nuclear power for 2030. The report also urges attention be given to the fact that nearly half of the DP participants changed their views between T1 and T2 as well as between T2 and T3, and that as many as a quarter of the DP participants continued to debate which option to choose even after two days of deliberation.

Third, the report discusses the overwhelming support for the zero scenario expressed by the public in their comments as well as by the applicants to the public hearings. Of course, the report reminds us that these results are less representative of public voice than those of opinion polls or the DP. Nevertheless, the report emphasized the significance of the surge in direct support for the zero scenario or the complete phase-out of nuclear power.

> As many as 77,000 public comments demanded a zero nuclear policy, and there are demonstrations every week against the restart of nuclear power stations. These reflect the public's distrust of the government and their concerns about nuclear safety. The first priority should be to overcome such distrust and concerns. ("For the formulation of the Strategy: Implications of the National Discussion" pp. 5-6)

This statement can be regarded as one of the most significant outcomes of the DP, and we will discuss this issue in Sect. 5.6.2.

Fourth, the report also discusses how to overcome conflicting opinions, listing eleven points of discussion extracted from the results of the National Discussion as follows.

1. Is it possible for the strengthening of safety measures to prevent a recurrence of the nuclear accident?
2. Nuclear power might turn out to be more expensive, considering health hazards and decontamination cost.
3. What to do with spent nuclear fuel? What are the realistic options and what responsibility should the government take?
4. It is not acceptable, on the other hand, to increase the amount of spent fuel when there is no clear plan for its disposal.
5. How to gather talented people to take charge of nuclear safety, including decommissioning projects, while the country is decreasing nuclear power dependency?
6. How is it possible to accelerate the development of renewable energy and energy conservation technologies? How can a stable energy supply be maintained when the development is stagnant?
7. Will renewable energy and energy conservation turn out to be cheaper than fossil fuel or nuclear energy? When will this be realized?
8. Won't there be a hollowing out of industry and job losses due to the increase in energy costs and instability of supply?

9. The development of renewable or new energy should be turned into opportunities for the creation of new industries and employment.
10. What should be the scenario after 2030?
11. How to encourage public participation in order to restore public trust in the government and public sector?

The report emphasizes that further discussion is necessary and that the government should express clear principles about (1) how to overcome the public's distrust of and concern about nuclear energy, (2) how to expand the use of renewable energy and energy conservation measures, and (3) their vision for a society built upon the current energy policy choice.

The final report was mostly planned and drafted by the secretariat, and the contribution of the Panel was limited to the provision of supplementary commentary on the given draft. Due to the time constraints, not all of the Panel members had enough time to prepare for the meeting with a careful reading of material on the results of the National Discussion. A considerable part of the Panel discussion involved no more than abstract, theoretical arguments, which should have been completed well before the National Discussion started. For all that, this one-week review process was not without value in that it was an unprecedented attempt to create a connection between the public participation process and policy decisions.

5.5.3 The Final Policy Decision and Its Aftermath

With the publication of the final report of the National Discussion, the formulation process of the energy strategy entered its final phase. In this phase, the prefectural and municipal governments in Aomori, where the nuclear fuel reprocessing plant and other nuclear-related facilities are located, made a strong protest against the zero nuclear option. Aomori has accepted spent nuclear fuel on the promise that the prefecture provides a site for the nuclear fuel cycle project and would never be turned into the final disposal site. The people of Aomori thought that Japan would one day withdraw from the nuclear fuel cycle project if the zero nuclear policy was adopted and that would place them in jeopardy. The power industry also tried to discourage the government from its target of zero nuclear power, enumerating a number of possible drawbacks to the policy.

Facing such a backlash in the final phase, the government began to consider the various risks associated with the zero nuclear option. In the course of the discussion, METI issued a memorandum on "The problems of the zero nuclear policy," which covered most of the issues raised by the concerned local governments and the power industry. The memorandum listed four challenges; that is, (1) obtaining the support of nuclear host regions, particularly in regard to spent nuclear fuel and the restart of existing nuclear power stations, (2) increased public burden due to zero nuclear power, (3) economic and security issues such as the influence on diplomacy and security or the loss of bargaining power in resource procurement, and (4) other mid- and long-term issues including the development of alternative energy sources, measures against global warming, and so on.

On September 14, the Energy and Environmental Council in the Cabinet finally decided the Innovative Strategy for Energy and the Environment, and the Strategy nevertheless targeted the realization of "a society that does not depend on nuclear power," "green energy revolution," and "stable energy supply." Based on these ideas, the Strategy announced three principles regarding nuclear power as follows:

1. Strict application of a forty-year limit to the operation of existing nuclear power plants
2. The Nuclear Regulation Authority (NRA)'s safety assurance as a requirement to restart the operation of nuclear power plants
3. No new or additional construction of nuclear power plants

The Strategy then declared, "The government will mobilize all possible policy resources to such a level as to even enable zero operation of nuclear power plants in the 2030s." Although the wording remained vague, and a few issues, including the future of the nuclear fuel cycle, remained unresolved, the National Discussion thus resulted in the decision to pursue "zero operation of nuclear power plants in the 2030s."

The DPJ coalition government intended to create a new national energy plan, but, in the general election in December 2012, it suffered a crushing defeat because of a number of policy missteps over the previous three years. As soon as the conservative LDP-led coalition took power, Prime Minister Shinzo Abe insisted that targeting a zero nuclear policy in the 2030s was unrealistic, and he declared "a zero-based review" of the Strategy, which in real terms meant its abandonment. This occurred just four months after the decision to pursue the zero nuclear strategy. In April 2014, the Abe administration decided a New Strategic Energy Plan. The plan regarded nuclear power as "an important base-load power source" that contributes to the "stability of energy supply-demand structure, on the major premise of ensuring of its safety," representing a major about-face on nuclear policy.

5.6 Discussion and Conclusion

5.6.1 The Quality of the National Discussion

One of the two main questions dealt with in this chapter is whether the Energy DP succeeded in allowing for substantial interactions between the citizen participants and those experts who had hitherto played more substantial roles in the decision-making process. The DP participants did more than just passively receive information from experts and vote on given policy options. Their questions extended over such nuanced topics as the individual preferences of the experts and the appropriateness of the policy options themselves. The categorization of the participants' questions that went beyond simply asking for technical information demonstrates that the questions covered three general areas, each of which can be thought essential to fundamental discussions.

First, there was a group of questions that challenged the information formally given in the briefing material. The questions on the safety or cost of nuclear energy asked on Day One are typical examples, and such questions helped reveal the various preferences of the experts.

Second, some participants questioned the experts' personal judgment regarding the options. In this context, an interesting example is the set of questions asked on Day Two in relation to whether the zero scenario was actually possible. These questions might appear somehow strange, considering the fact that the zero scenario was included among the policy options as most of the experts engaged in this energy debate agreed it was more or less possible. A large number of the DP participants thought that the zero scenario was the most desirable means to achieve a safe energy policy but, at the same time, they understood that there were great difficulties in terms of other criteria, such as economic efficiency and energy security. The participants were determined to decide for themselves, not leaving it solely to the experts, nevertheless they wanted to ascertain how realistic the individual experts considered the zero scenario to be for reference purposes.

Third, participants challenged the given policy options, asking whether there weren't any other options or how the given options were formulated. Participants also asked why the options included continued nuclear energy dependence despite the Fukushima disaster. These questions can be seen as the most radical as they challenge the very premise of the discussion. The experts found it difficult to respond to these questions as they themselves took the premises or framework of the discussion for granted.

Unfortunately, questions of this type did not elicit straightforward responses in the Energy DP. It was not that the experts clearly avoided answering such questions; rather they took a more subtle approach. In the plenary sessions, most questions in this category were combined with other technical questions and were not dealt with individually. As a result, the experts' answers mainly focused on technical details, which were naturally easier for them to answer, and in-depth questions challenging the framework of the discussion tended to be ignored. Due to time constraints, the DP program is not designed to allow participants to ask follow-up questions. If the participants had had opportunities for follow-up questions, the experts might have been forced to answer these questions more explicitly.

Although some of the in-depth questions were not answered openly, the significance of these questions was not lost. By posing the three types of questions described above, participants were not just asking for technical information, instead they questioned, even challenged, the validity of the prevailing official information, the experts' personal judgment on policy options, and the legitimacy of the discussion framework. Therefore, what mattered was not only the content of experts' answers but also the ways in which the experts presented the answers, particularly the fact that some experts avoided answering, intentionally or not, specific questions.

The results from the participant evaluation described above (Sect. 5.4.3) exhibit that participants were no less informed by the discussion points raised by other groups than by the responses given by the invited experts. As far as the observation of the DP discussion is concerned, the National Discussion succeeded in providing opportunities for substantial deliberation, not merely the one-way transfer of information from experts to the public.

5.6.2 Public Participation and Decision-Making

This chapter has also focused on the extent to which DPs or mini-publics can influence decision-making in the field of science and technology policy. In most cases, the connection between the public participation process and decision-making is obscured to outside observers, with only the policymakers themselves aware of its value. Lacking a substantial link to policy decisions, participation can be tokenistic or present unrealistic promises of more democratic governance (Felt and Fochler 2010).

In this sense, the importance of the review process cannot be overestimated although there were certain limitations in connection with the Review Panel itself. The disclosure of this post-participation process allowed an investigation of the way in which the results of the public participation were used and the degree to which they actually impacted policy. Any examination of the impact of the National Discussion, particularly that of the DP, on decision-making, should start with the conclusions (implications) of the review report.

First of all, the conclusion drawn in the review report that more than a half of the population would like to move away from dependency on nuclear energy can be said to reflect the DP results to a large extent. Approximately 90 % of public comments and 70 % of the opinions presented on the applications for participation in the public hearings demonstrated support for the zero scenario. However, the results from these conventional public participation methods are often thought to reflect only a narrow cross-section of public opinion. The great difference this time was that the results from the mini-public, the Energy DP, were also available. The support for the zero scenario increased as the DP participants gained more information and had greater opportunity for deliberation, with nearly half of the participants eventually supporting the zero scenario at T3. In addition, the analysis of T3 data revealed that approximately half of the supporters of the 15 scenario were in favor of nuclear phase-out at some time in the future. Thus, the conclusion that more than half of the population is in favor of a break from dependency on nuclear energy is based on solid evidence from the DP.

Another significant impact of the DP can be found in the review report's reference to the underlying cause of the "direct action" taken by the public, such as the large number of public comments and demonstrations calling for a zero nuclear option. The State Minister in charge of the National Strategy published the review report, and it is not surprising that he should have mentioned the public comments, which appeared in response to an official process. It is surprising, and unusual, however, that anti-nuclear demonstrations, quite clearly 'uninvited' participation (Wynne 2007), were referred to in such a positive manner in this formal document.

It is a well-known episode in Japanese contemporary history that the former Prime Minister, Nobusuke Kishi, once stated, "I have to listen to the 'silent majority (*koe naki koe*).' All I can hear now is the 'loud minority (*koe aru koe*).'" This occurred in 1960, when he was surrounded by demonstrators protesting against the Japan-US Security Treaty. This famous phrase was given in a press conference, and Kishi continued by referring to the demonstration outside the Diet as the "loud minority," saying, "Yielding to this 'loud minority' would place Japan in crisis."[18]

Fifty-two years later, in June 2012, the Japanese public was reminded of the comments made by the late Prime Minister Kishi when the then Prime Minister Yoshihiko Noda remarked to reporters with regard to a crowd of demonstrators, "They're making lots of noise."[19] Mr. Noda was criticized for interpreting the voice of protest as mere "noise." Although he might not have meant to be offensive, this episode seemed to reveal that politicians' view of the demonstration as a "loud noise" drowning out the "silent majority" remained.

Despite the obstinate belief that direct public action is representative of only the "loud minority," the government finally concluded in the review report, "[The public comments and demonstrations requesting a zero nuclear policy] reflect people's distrust of the government and concerns about nuclear safety." It is reasonable to consider that the Energy DP, by highlighting the outrage and distrust felt by the "silent majority," provided a significant background to this historic conclusion. The Energy DP showed that at least half of the population supported a zero nuclear option with the hope of making safety a first priority, and this result compelled policymakers to regard the public comments and demonstrations as something akin to the public sentiment displayed in the DP, rather than as a "noisy minority."

Nevertheless the about-face after the 2012 general election demonstrated the vulnerability of public participation and deliberation to changes in political circumstances. The LDP-led coalition government did not merely reject the energy strategy formulated by the DPJ administration, but ignored the results from the National Discussion including the DP. Further discussion on a national scale on a number of unsolved questions, particularly regarding the restart of nuclear power stations, the future of nuclear fuel cycle policy, and the disposal of high-level radioactive waste, was neglected. Thus, the National Discussion can be seen to have been no more than a small ray of hope among a mountain of challenges facing participatory and deliberative nuclear governance in this country.

---

[18] The Asahi Shimbun, May 28, 1960 (evening edition).

[19] Mr. Noda's comment was in Japanese (he was reported to have said, "*Okina oto ga shimasune.*"), and the English translation shown here in the text is from the New York Times, June 29, 2012. http://www.nytimes.com/2012/06/30/world/asia/thousands-in-tokyo-protest-the-restarting-of-a-nuclear-plant.html. Accessed 20 Aug 2014.

5.6.3 Conclusion

The essence of mini-public-type public participation is to provide opportunities for deliberation on important public issues to ordinary members of the public who are not necessarily interested in the issues at hand. Through the process, many participants are encouraged to think about the issues and the information provided, questioning and challenging the premises and framework of the discussion as well as the appropriateness of the policy options themselves. Once reported through the media, the deliberative process in a mini-public gives other members of society an opportunity to think about and discuss the issues.

The deliberation process appears to continue within and around the individual participants even after the mini-public forum. Although the author did not have access to the actual participants of the Energy DP, he was able to interview some of the citizen panelists who had participated in a public forum using a modified DP method, organized in parallel with the government-sponsored DP.[20]

One of the interviewees, a female office worker, reflected on the change that she experienced after the forum, saying:

> At first, I tended to think that economic reform is more important than energy and nuclear policy. However, I came to understand there are various energy policy options, including a zero nuclear policy. Even though I didn't change my attitude greatly at the time of post-event survey [T3], I have gradually changed my mind toward zero nuclear dependency while watching TV broadcasts on other countries that stopped using nuclear power. (Interviewed on November 14, 2012, by telephone)

The point here is not that this participant changed her attitude from a pro-nuclear stance to support for the zero option. Rather, the important point is the fact that she continued to deliberate on the issue even several months after the forum. On the day of the forum, a TV crew kept up close coverage of her group, and she happened to appear in a news program. One of her coworkers noticed her on the broadcast, and later spoke to her about it. That led them to discuss nuclear energy issues, which had seldom been a topic of conversation at their office in the past. In relation to this, the participant explained, "The participation in the forum actually led me to a wider discussion." This is only one episode from the independent public forum, but it is not unlikely that the 285 participants of the Energy DP had similar experiences after their involvement in the forum.

The National Discussion revealed the possibility of emphasizing the voices of the "silent majority," not as an aggregation of reflex responses to opinion polls but as the voice of real flesh-and-blood people. Although this possibility is still

---

[20] The forum recruited the participants from 3,000 randomly selected residents of Kawasaki-city, Kanagawa Prefecture. A total of 670 people responded to T1, and 57 people eventually participated in the discussion forum held in August 2012 (Miyagi and Yagishita 2013; Yagishita (Ed.) 2014, pp. 50–51). The author was a member of the organizing committee of this project and interviewed with four ex-participants from November to December in 2012.

particularly vulnerable to the chaotic world of real politics, it is necessary to continue searching for room in Japanese society for such voices to be heard, and that should be one of the essential lessons from Fukushima.

Acknowledgments The work of this chapter was supported by JSPS KAKENHI Grant Number 24501085, 26340111.

References

Blok, A. (2007). Experts on public trial: On democratizing expertise through a Danish consensus conference. *Public Understanding of Science, 16*(2), 163–182. doi:10.1177/0963662507062469.
DP Steering Committee (Enerugi kankyo no sentakushi ni kansuru torongata yoronchosa jikkoiinkai) (2012a). Enerugi kankyo no sentakushi ni kansuru torongata yoronchosa toronshiryo (Briefing booklet for the Deliberative Polling on Energy and Environmental Policy Options).
DP Steering Committee (Enerugi kankyo no sentakushi ni kansuru torongata yoronchosa jikkoiinkai) (2012b). Enerugi kankyou no sentakushi ni kansuru torongata yoronchosa chosa hokokusho (Report on the Deliberative Polling on Energy and Environmental Policy Options).
Dryzek, J. S., Goodin, R. E., Tucker, A., & Reber, B. (2009). Promethean elites encounter precautionary publics: The case of GM foods. *Science, Technology and Human Values, 34*(3), 263–288. doi:10.1177/0162243907310297.
Einsiedel, E. F., Jelsøe, E., & Breck, T. (2001). Publics at the technology table: The consensus conference in Denmark, Canada, and Australia. *Public Understanding of Science, 10*(1), 83–98. doi:10.1088/0963-6625/10/1/306.
Felt, U., & Fochler, M. (2010). Machineries for making publics: Inscribing and de-scribing publics in public engagement. *Minerva, 48*(3), 219–238. doi:10.1007/s11024-010-9155-x.
Fishkin, J. S. (2009). *When the people speak: Deliberative democracy and public consultation.* New York: Oxford University Press.
Goodin, R. E., & Dryzek, J. S. (2006). Deliberative impacts: The macro-political uptake of mini-publics. *Politics and Society, 34*(2), 219–244. doi:10.1177/0032329206288152.
Hasegawa, K. (2014). The Fukushima nuclear accident and Japan's civil society: Context, reactions, and policy impacts. *International Sociology, 29*(4), 283–301. doi:10.1177/0268580914535413.
Honda, H. (2005). *Datsu genshiryoku no undo to seiji: Nihon no enerugi seisaku no tenkan wa kano ka (Movements and politics of post-nuclear power).* Sapporo: Hokkaido University Press.
Irwin, A. (2006). The politics of talk: Coming to terms with the 'New' scientific governance. *Social Studies of Science, 36*(2), 299–320. doi:10.1177/0306312706053350.
Kobayashi, T. (2004). *Dare ga kagaku-gijutsu ni tsuite kangaerunoka: konsensasu kaigi toiu jikken (Who should deliberate on science and technology? Consensus conference as an experiment).* Nagoya: Nagoya University Press.
Kobayashi, T. (2012). "Kokuminteki giron" towa nandattanoka: Genpatsu o meguru shiminsanka no arikata (What was the "National Discussion"? Public participation on nuclear power). *Asteion, 77*, 192–208.
Miyagi, T. (2014). Kakushinteki enerugi kankyo senryaku no seisaku kettei katei (The policymaking process of the Innovative Strategy for Energy and the Environment). In M. Yagishita (Ed.), *Tettei togi nihon no enerugi kankyo senryaku (Public debate on the Innovative Strategy for Energy and the Environment)* (pp. 1–26). Tokyo: Sophia University Press.
Miyagi, T., & Yagishita, M. (2013). Torongata yoronchosa no shuho o mochiita minkan dokuji chosa no kokoromi: 3.11 go no enerugi-kankyo no sentakushi ni kansuru kokuminteki giron (A private sector-led study of public opinion using the deliberative poll method: Public debate

on the Innovative Strategy for Energy and the Environment after March 11, 2011). *Chikyu kankyo gaku (Global environmental studies), 8*, 79–112.
Nielsen, A. P., Lassen, J., & Sandøe, P. (2007). Democracy at its best? The consensus conference in a cross-national perspective. *Journal of Agricultural and Environmental Ethics, 20*(1), 13–35. doi:10.1007/s10806-006-9018-5.
Nishioka, S. (2011). *Teitanso shakai no dezain: Zero haishutsu wa kano ka (Designing a low-carbon society: Is zero-emission possible?)*. Tokyo: Iwanami Shoten.
Ogiwara, S. (2014). "Enerugi kankyo no sentakushi ni kansuru torongata yoronchosa" ni kansuru shimbun kiji no doko (A review of newspaper articles on "Deliberative Polling on options for Energy and Environmental Policy Options"). *Jinbunkagaku nempo, 44*, 25–43.
Onai, T. (2007). Nihon ni okeru "jukugi-sanka demokurashi" no hoga: Genshiryoku seijikatei o toshite (The emergence of "deliberative-participatory democracy" in Japan: Through nuclear politics). In A. Ogawa (Ed.), *Posuto daihyosei no hikakuseiji: Jukugi to sanka no demokurashi (Comparative politics of the post-representative era: Democracy of deliberation and participation)* (pp. 79–104). Tokyo: Waseda University Press.
Onai, T. (2014). Jukugi minshushugi (Deliberative democracy). In H. Honda & T. Horie (Eds.), *Datsu genpatsu no hikakuseijigaku (Comparative politics of post-nuclear power)* (pp. 109–128). Tokyo: Hosei University Press.
Powell, M., & Kleinman, D. L. (2008). Building citizen capacities for participation in nanotechnology decision-making: The democratic virtues of the consensus conference model. *Public Understanding of Science, 17*(3), 329–348. doi:10.1177/0963662506068000.
Shinohara, H. (Ed.). (2012). *Togi demokurashi no chosen: mini paburikkusu ga hiraku atarashii seiji (The challenges of deliberative democracy: Democratic innovation through mini-publics)*. Tokyo: Iwanami Shoten.
Smith, G. (2009). *Democratic innovations: Designing institutions for citizen participation*. New York: Cambridge University Press.
Sone, Y., Yanase, N., Uekihara, H., & Shimada, K. (2013). *"Manabu, kangaeru, hanashiau" Torongata yoronchosa: Giron no atarashii shikumi (Deliberative Polling "to learn, think, and talk": A new system for discussion)*. Tokyo: Kirakusha.
Sugawara, T. (2013). Torongata yoronchosa (DP) o kangaeru: Chosa kenkyukai hokoku (Discussing Deliberative Polling (DP): Research workshop report). *Bulletin of the Japan Association for Public Opinion Research, 111*, 60–69.
Takahashi, H. (2011). *Denryoku jiyuka: hassoden bunri kara hajimaru nihon no saisei (Electric power deregulation: Revitalization of Japanese society through the separation of power generation and transmission)*. Tokyo: Nihonkeizai Shimbun Shuppansha.
Wakamatsu, Y. (2010). *Kagaku-gijutu seisaku ni shimin no koe o do todokeruka: konsensasu kaigi, shinario wakushoppu, dipu-daiarogu (How to deliver the voice of the public to science and technology policy? Consensus conferences, Scenario Workshops, and Deep Dialogues)*. Tokyo: Tokyo Denki University Press.
Wynne, B. (2007). Public participation in science and technology: Performing and obscuring a political-conceptual category mistake. *East Asian Science, Technology and Society: An International Journal, 1*(1), 99–110. doi:10.1007/s12280-007-9004-7.
Yagi, E. (2013). Enerugi seisaku ni okeru kokuminteki giron towa nandattanoka (What was the National Discussion on Energy Policy?). *Journal of the Atomic Energy Society of Japan, 55*(1), 29–34.
Yagishita, M. (Ed.). (2014). *Tettei togi nihon no enerugi kankyo senryaku (Public debate on the Innovative Strategy for Energy and the Environment)*. Tokyo: Sophia University Press.
Yanase, N. (2013). Kokyoseisaku no keisei eno minshuteki togi no ba no jisso: Enerugi-kankyo no sentakushi ni kansuru torongata yoronchosa no jisshi no gaikyo (The implementation of the deliberative forum for public policy making: Report on the National Deliberative Polling on Energy and Environmental Policy Options). *Komazawa daigaku hogakubu kenkyu kiyo, 71*, 53–186.
Yoshioka, H. (2011). *Shimpan Genshiryoku no shakaishi: Sono Nihonteki tenkai (Social history of nuclear energy: The case of Japanese development)*. Tokyo: Asahi Shimbun Shuppan.

Part II
Historical Construction of Science, Technology and Society Relationship

Chapter 6
Minamata Disease: Interaction Between Government, Scientists, and Media

Shigeo Sugiyama

Abstract It is often said that the Great East Japan Earthquake and the Fukushima Daiichi Nuclear Power Plant accident shook people's trust in science. This suggests that people had, until then, unreasonable or unrealistic expectations about what science can achieve and how "scientific knowledge" functions in society. In fact, in the Minamata Disease Incident, which occurred during the late 1950s, a solution was hampered by people's incorrect understanding of the realities of science, along with a failure to provide people with a correct understanding. (1) In science, even if certain things are yet not fully understood, it is still possible to state that "We know this much at least." This point should have been clarified. (2) The nature of scientific explications is such that they are gradually refined and made more accurate using a range of theories. As this process was not highlighted along with the reasons for this need for refinement, it gave the impression that scientists were merely running about in confusion. (3) A vicious cycle resulted, wherein the more scientists endeavored to conduct a thorough scientific inquiry into the causes of Minamata disease, the less action the Japanese government took on the grounds that they had yet to reach a definite conclusion. While science, in general, is the pursuit of truth that transcends temporal restrictions, scientists need to be wary of colluding with policy makers when research results are expected for social policies.

6.1 Introduction: Events that Occurred in Minamata

"It's cerebral palsy" diagnosed a private clinic's medical practitioner, talking to the mother of a 7-year-old patient. However, at a municipal hospital pediatrics department, which they had visited the previous day, the mother was told that her daughter was malnourished. At yet another private clinic that they had visited five days before this, the diagnosis was infantile paralysis.

S. Sugiyama (✉)
Hokkaido University, Sapporo, Japan
e-mail: sugiyama.shigeo@nifty.com

Certainly, it was a mysterious, unidentified disease that this little girl was suffering from. As became apparent later, several other patients suffered from the same mysterious symptoms. Moreover, a number of cats in these patients' houses or neighborhoods had suffered convulsions and died.

A few weeks later, the outbreak of a "strange infant disease" was reported to the health center in Minamata. It was called "strange disease" because it was a previously unknown illness with an unknown cause. Thus, the disease later called "Minamata disease" came to light in May 1956.

As soon as patients with this strange disease were officially recognized, researchers at the School of Medicine, Kumamoto University made every effort to elucidate its nature and cause. At first, the disease was believed to be infectious. However, because no virus was detected, this notion was quickly dismissed. By the end of 1959, the strange disease was believed to be poisoning by some kind of heavy metal.[1]

It proved difficult, however, to pinpoint this heavy metal. Between the end of 1956 and early 1957, rumors began to circulate that the metal was manganese. On February 14, 1957, the *Kumamoto Nichinichi Shimbun* ran the headline, "Strange Disease in Minamata: Is Manganese the cause?" Two months later, however, on April 17, the same newspaper attributed it to selenium with the headline, "Is it Selenium poisoning? School of Medicine Kumamoto University issues warnings over strange disease in Minamata." Not long after this, thallium became a suspect.

During the many twists and turns in this saga, in July 1959, organic mercury surfaced as a suspected substance. Its origin was inferred to be industrial wastewater from the New Japan Nitrogenous Fertilizer Company (NJNFC) (hereafter the factory). On July 16, the *Kumamoto Nichinichi Shimbun* reported the research group's conclusions under the headline, "Minamata disease/caused by organic mercury/conclusion drawn by Kumamoto University Research Group." On July 22, the newspaper announced that a conference would be called, inviting representatives from the municipal assembly and a hospital attached to NJNFC. On July 23, the newspaper reported on this briefing session under the headline, "Organic mercury poisoning/cause of Minamata disease/detected in urine and seafood/result announced unanimously by Kumamoto University Research Group/'Conclusion almost certain.'" The main text of the article ran as follows:

> At the briefing session, all researchers were unanimous in their conclusion that the cause of Minamata disease is almost certainly organic mercury, and it is of great significance that Minamata disease, which has remained a mystery illness since its outbreak in 1953, has finally shown its true nature.

---

[1] For example, on November 26, 1956, the *Kumamoto Nichinichi Shimbun* ran an article with the headline, "Mystery Illness in Minamata not contagious/no virus discovered/focus of investigations on poisoning/Kumamoto University Professors hold Press Conference."

Thus, there was a great deal of elation that the cause of Minamata disease had finally been revealed. A newspaper article from July 15 reported,

> Now that the cause of Minamata disease has been revealed, urgent action can be taken to put future measures in place, such as special legislation for no-fishing zones and providing compensation to those affected and fishermen.

The general feeling was that an overall solution was not far away.

Before long, the factory began to refute these claims. On August 5, at a special prefectural committee meeting about Minamata disease countermeasures, factory representatives pointed out several issues, claiming: "[the organic mercury theory] has no validity and is nothing more than an inference; moreover, linking this with wastewater from the factory is an illogical leap" (*Kumamoto Nichinichi Shimbun*, August 6). Following this, the factory continued to issue a series of counterarguments, while simultaneously altering the details. Its four main points of contention were as follows (Minamata Factory 1959):

1. Since before the war, the factory has been using mercury to produce acetic acid and vinyl chloride, and it is a fact that part of it was transmitted to Minamata Bay. Meanwhile, the fishing methods of fishermen and diets of residents near Minamata Bay have remained unchanged for the past several decades. In that case, why would mercury suddenly cause Minamata disease now? This does not make sense.
2. Acetic acid and vinyl chloride have long been produced at factories around the world, using the same methods as the ones used in the Minamata factory. Japan alone has over 15 such factories, over half of which are located along coasts. If the cause of Minamata disease were mercury, then Minamata disease should also be occurring in other places, but this is not the case. This does not make sense either.[2]
3. The mercury in industrial wastewater is inorganic. If organic mercury originating from industrial wastewater were responsible for causing Minamata disease, there would be a need to provide proof through scientifically accurate empirical data of the mechanisms by which inorganic mercury turns into organic mercury.
4. Organic mercury is known for dissolving in organic solvents such as alcohol. And yet, it has been confirmed through experiments that the toxic substances found in seafood do not dissolve even if treated with organic solvents. There is a clear contradiction here.

The repeated assertion of these four persuasive counterarguments[3] led to a sudden reversal in the belief that the cause of Minamata disease would have been

---

[2] It is common sense that the factory, considering the abovementioned (1) and (2), thought that some kind of change took place in Minamata Bay between 1953 and 1954, pointing the finger of suspicion at munitions dumped by the Japanese army in Minamata Bay shortly after the Second World War. The Japan Chemical Industry Association supported this munitions theory.

[3] The counterargument (4) by the factory "was not noticed until Moore pointed it out in 1962, and was considered as quite strong counterevidence against the organic mercury theory" (Tomita and Ui 1969). The Kumamoto University researchers also had difficulty explaining (3). Even in "Observations on the cause of Minamata disease viewed mainly from a Pathological Perspective" (July 22, 1959), they are hard-pressed to put forward an explanation. Any doubts were eventually dispelled when direct evidence was obtained that organic mercury was originally present in industrial wastewater (announced in February 1963). A proper explanation was given for (1) and (2) 40 years later (Nishimura and Okamoto 2001). Therefore, as of 1959, the factory's

found soon and that an overall solution was not far away. For instance, just two days after the NJNFC first put forward its counterarguments, the *Kumamoto Nichinichi Shimbun* printed the following on August 7.

> There is once again uncertainty surrounding the cause of Minamata disease. In response to the organic mercury theory posited by the Kumamoto University research group, a team of scientists from the NJNFC Minamata Plant has put forward a counterargument, stating that this is irresponsible conjecture. With images of industrial wastewater in mind, we too thought, "So that was it?" But of course, we had tacitly suspected that this was the cause.

Against this background, in November 1959, the Food Sanitation Investigation Council filed a report in response to an enquiry from the Minister of Health and Welfare. Here, despite stating that "the cause of Minamata disease is some kind of organic mercury compound," the existence of persuasive counterarguments (i.e., some points which remain unaccounted for) led to the decision that "further research was needed," thus leaving the matter unsettled.[4]

While the factory strongly refuted the organic mercury theory, Hajime Hosokawa, a doctor at the factory's hospital, felt that there was something not quite right about the factory's counterarguments. Therefore, wanting to know whether the factory was responsible, he initiated his own experiments, in which he gave the suspected industrial wastewater to cats. He believed that if the cats did not exhibit Minamata disease symptoms, then the factory's industrial wastewater was not responsible.

Hosokawa conducted his experiments without the company's knowledge. He had a subordinate bring him some wastewater discharged during acetaldehyde processing and mixed 2 cc with some food every day. He then fed the wastewater-laced food to the cats. On October 6, Cat 400 began to exhibit strange symptoms. Hosokawa wrote in his notes as follows:

"The cat is listless and curled up. It staggers around in its cage a bit. It has an appetite. When allowed to walk outside, mild paralysis of the hind legs can be observed. Its fur has lost all luster." The cat's symptoms progressed, until, on October 24, it ran around in circles on two occasions. "It has gradually weakened, become completely listless, and lost all appetite (Hosokawa 1959)." At the hospital, whether cats exhibited symptoms of Minamata disease was judged on the basis of whether the cat "ran around in circles." On this occasion, it did run in circles. A report detailing the results of a pathological examination, returned from Kyushu University in mid-November, listed findings characteristic of Minamata disease, such as a drop or loss of granule cells in the cerebellum.

However, Tadashi Ichikawa, deputy manager of the technical department of the factory, who was beginning to find out that Hosokawa was conducting experiments,

---

(Footnote 3 continued)
counterarguments contained content that could not be ignored, which is why in this sense they were "persuasive."

[4] The course of events during this period is described in detail in a number of books such as Ui (1968).

heard of Cat 400's symptoms and told Hosokawa not to disclose his findings to anyone. His reason was "a single example does not constitute proof." He said that unless Hosokawa could confirm the link through a suitable number of case studies, it was not possible to say that (something in) the wastewater had actually caused the symptoms of Minamata disease; some other factor might have been at play. He also suggested, "Let's go public after we have conducted further tests and are absolutely sure (NHK Special and NHK news crew 1995)."

Hosokawa eventually accepted Ichikawa's advice and continued to maintain silence about Cat 400, even after he had resigned from the company. He went public about this incident only much later, when he was called to testify at the Minamata disease trials in 1970.

Not long after the Kumamoto University research group had proposed the organic mercury theory, the group received, between the fall of 1959 and early 1960, criticism for repeatedly changing its opinions.

Mizu, a trade magazine related to water treatment, raised this point in its January and February 1960 issues. This magazine lambasted the Kumamoto University School of Medicine research group, arguing, "While [the School of Medicine] pointed suspicions to manganese at the end of 1956, by 1958, in the three years since its original statement, it changed its conclusion three times—to selenium, thallium, and, by July 1959, had developed the theory of 'organic mercury poisoning.'" The magazine further added that the group's "theories lack consistency, being revised often with no guiding principles" and branded Kumamoto University as a "third-rate provincial university." Furthermore, in its October 15, 1959 issue, the *Minamata Jiji Shinpo*, a printed mimeograph distributed to local officials and businesses, wrote, "Academic clique or just plain fickleness—they have changed the cause four times now." In September, the factory also criticized the group, suggesting, "The suspected substance has changed four times since 1956: manganese, selenium, thallium, and organic mercury; moreover, perhaps they issued their statements at any time based on the judgment that the clinical and pathological findings for each substance resembled Minamata disease."

6.2 Overlooked Points of Contention

After beginning in such circumstances, the Minamata Disease Incident has yet to be completely resolved despite nearly 60 years having passed (at this writing) since patients were first officially recognized in 1956. Many patients still need help, and certain aspects of responsibility by the Japanese government and Kumamoto Prefecture's remain unresolved. Thus, the Minamata Disease Incident illustrates how governance over scientific technology failed. This failure was one of the largest of its kind, both in terms of the immensity of damage and the length of time needed to reach a resolution.

In fact, there were numerous opportunities to discover the correct means to a resolution: when it became clear that Minamata disease was a food poisoning that occurred through consumption of seafood caught in Minamata Bay (1956–1957)[5]; when all fingers pointed to organic mercury originating from the factory's industrial wastewater (summer–fall 1959); when the Japanese government officially recognized methyl mercury in the factory's industrial wastewater as the cause of Minamata disease (September 1968). Despite these opportunities, even today the Minamata Disease Incident remains largely unresolved. But what, exactly, brought about such a bewildering situation?

To date, three overall issues have been highlighted: For instance, the factory's fervent and active attempt to avoid any responsibility: In September 1958, when scientists from Kumamoto University were investigating toxic substances in seafood, American medical scientist Dr. Leonard T. Kurland and others visited Minamata to conduct surveys. In Japan, Dr. Kurland requested samples of local seafood to take back to the United States for examination. However, when the factory learnt of this, they quickly purchased all the fish caught in Minamata Bay (In the event, Kurland and others took back samples taken by Minamata Health Center).

The lack of action by Kumamoto Prefecture and the national government have also been highlighted. NJNFC's purchase of all the fish is just one of six incidents described in a Kumamoto Prefecture sanitation section document titled "Examples of the New Japan Nitrogenous Fertilizer Company's Uncooperativeness in Researching the Cause of Minamata disease (October 1959)." The document has even been stamped with a "classified" seal. In other words, although prefectural authorities were aware of NJNFC's problematic actions—at the very least, in its response to determining the disease's cause—it concealed the truth through inaction. We do not go into detail here, but this was also the case with the Ministry of International Trade and Industry as well as the Ministry of Welfare.

Third, the factory's position within the local community has been highlighted. Minamata was the company's headquarters, with its citizens depending on the company for their livelihoods. Moreover, many Minamata disease patients were from poor, outlying fishing villages, illustrating a feeling of discrimination toward them among residents.

But were there no other factors or problems worthy of consideration? Were there no problems with people's understanding of science (i.e., the nature of science and its functions in society) or forms of communication (such as methods of publishing scientific information, ways of comprehending published scientific information)? Next, we consider these points based on events that occurred during the early stages of the Minamata Disease Incident (1956–1959), described in Sect. 6.1: Introduction: Events that occurred in Minamata.[6]

---

[5] Tsuda (2004) has given a detailed account from this perspective.

[6] If the results of the Cat 400 experiment had been publicly announced in 1959, it is highly likely that the course of the Minamata Disease Incident would have taken a different direction, which would have reduced the number of victims. For a more detailed account of the Cat 400 experiment, see Sugiyama (2005). This paper uses the results of the Cat 400 experiment as an "example of failure," in that it did not lead to an early resolution of the Minamata Disease Incident, to show

To repeat, this paper's purpose is not to suggest that the following points independently form the essence of the Minamata disease incident, or that they are the most significant issues.[7] While little mention has been made of these points, we argue that they are factors behind the Minamata Disease Incident, and that even today, they have yet to be overcome.[8]

6.3 Reconsidering Occurrences in Minamata

6.3.1 Counteractions

As described in Sect. 6.1, from the summer of 1959, the New Japan Nitrogenous Fertilizer Company, Kumamoto Prefecture, and the Japanese government acted to nullify the organic mercury theory and to legitimize inaction. Jun Ui, environmental scientist, compared the expression of such "counteractions" to acids neutralized by alkalis: launching a series of prominent counterarguments against a convincing theory gradually nullified its overall persuasiveness. He observed this as a common process in many regions that have experienced pollution incidents (Ui 1968; Tomita and Ui 1969).

However, this process is not limited to pollution incidents, but occurs frequently when science and technology become a major social issue, as in the case of global warming. Therefore, to move forward from this situation, we must question why counteractions were so effective in the Minamata Disease Incident.

The core of counteractions lies in lowering the overall credibility of the other party's theory by criticizing their arguments. This kind of counteractions is believed to be effective when ideas like the following have gained widespread currency:

> 'Solved scientifically' means that something has been explained perfectly in all respects; thus, it is impossible to say that something has been successfully solved scientifically if some aspects remain unexplainable (problems remain).

In other words, if a theory contains some facts that are inexplicable, then the theory's legitimacy can be denied in its entirety. Certainly, the factory refuted the organic mercury theory by highlighting its flaws: (i) some facts had yet to be explained (inexplicable aspects), and (ii) certain facts contradicted this theory (and

---

(Footnote 6 continued)
how no matter how scientifically uncertain, it should not be confined to the decision making of professional scientists (some kind of public decision making is required).

[7] The author believes that things not visible from previous perspectives have become visible from new perspectives, and that they contain universal lessons that transcend the individual case of the Minamata Disease Incident.

[8] For example, when the factory's handling of the situation (counterarguments) were shown to young people today, over half felt that its way of handling the situation and counterarguments were reasonable.

that which it predicted). Thus, the factory attempted to destroy the theory's credibility through its flaws.

The understanding of what "solved scientifically" means as described above seems to be an idea commonly accepted by many people. For example, a group of people were asked to read Masahiko Fujiwara's essay, "I Turn my Bowl but Green Onions Don't Come Around." Thereafter, they were asked what they thought about his final sentence, "The scary thing about science is even the most plausible theories can be blown away by a single counterexample." Over half of the respondents answered "I completely agree" or "If I have to choose, then yes."[9]

Science studies of a while ago often discussed that science is incomplete when aspects of phenomena have yet to be explained or there are facts that are thought to contradict theories. Nevertheless, this "incompleteness" is a standard part of actual science. Even if scientists encounter facts that contradict a theory, they seldom discard the theory completely. Instead, they try to explain the facts using the theory in question, reviewing the authenticity of those aspects that are considered as facts. By maintaining the theory's core and revising some of its peripheral aspects, they aim to determine whether they can explain the facts (i.e., whether they can elucidate the effect of these facts on the functioning of the theory). And in many cases, they succeed in doing so. In other words, theories are not hard structures that can be disregarded merely due to the existence of a few contradictory facts.

In actual scientific research, of course, it is typical for certain theories to be clearly refuted or affirmed through accurately planned and conducted experiments. Fujiwara's essay "I Turn my Bowl but Green Onions Don't Come Around" models itself on such cases. In terms of number, such cases can be said to be much more common than the other way around.[10] However, with increasingly complex phenomena, clearly determining whether something is right or wrong is more difficult. And, of those phenomena that become the subject of social controversy—the science that the general public encounters—the overwhelming majority are cases such as these. Therefore, people's scientific understanding should align with scientific realities that they are likely to encounter.

Indeed, the very nature of scientific endeavor is the persistent and ongoing pursuit of truth. From this perspective as well, science is imperfect by its very

---

[9] This essay can be summarized as follows (*Asahi Shimbun*, August 10, 2002 evening edition). The author finishes eating a bowl of ramen (noodle soup) and notices that pieces of green onion were floating on the soup's surface. He tries to turn the bowl around to bring them in front of him but without much success. His eldest son tells him the reason for this: "It is because ramen has a lot of oil and broth slides over the bowl's surface." He then says that he will try it with soba noodles, which do not have oil. Upon which, his third son replies that, "The same thing would happen even with ice floating in a cup of water: you would turn the cup and it wouldn't move." This served to blow away the oil theory. This episode is followed by the sentence, "The scary thing about science is even the most plausible theories can be blown away by a single counterexample."

Thinking about it, school science education is also based on this kind of idea. Experiments which appear on television programs also serve to prove or disprove a certain theory in a simple manner, thus strengthening such generally accepted notions.

[10] In this sense, the way it is dealt with in regular science education is not complete nonsense.

nature. Even assuming that scientists deem knowledge about a specific phenomenon has "reached a stage of completion," they subsequently ask new questions or pose new problems in pursuit of even deeper understanding. In practice, scientists are never satisfied with their current understanding, and in this sense, the normal state of science is that it is still incomplete.

However, "There are still parts we do not understand" does not mean "We do not understand anything yet." "There are some things that we do not yet understand" implies, in other words, that "many aspects have been elucidated." In some cases, it is even possible to assert that "the essential parts have already been understood."

Scientific research can be compared to a crossword puzzle in which the horizontal and vertical clues help determine a word in certain bolded squares. Once some bolded squares are filled in and their letters lined up, the target word can often be deduced without completing all the squares, i.e., without solving the entire puzzle.

This kind of situation certainly occurred in the Minamata Disease Incident. Kumamoto Prefecture instructed fishermen to cease fishing voluntarily and then conducted trials in pearl cultivation to encourage the fishermen to change their occupation. This can only mean that prefectural authorities already knew the cause of Minamata disease. It was pointed out long before the notion of organic mercury in industrial wastewater surfaced that "what is most suspected currently is poisoning due to the consumption of seafood caught in Minamata Bay (Kousei Kagaku Kenkyu Han 1957)," and this was common understanding among those concerned. Neither the government, the prefecture, nor the factory objected. Despite this, they asserted that "We know nothing yet," using the rationalization that "Kumamoto University's organic mercury theory is still incomplete."[11]

Therefore, to prevent such counteractions, the following points are vital:

(a) There must be an understanding that, even if some parts of a phenomenon are not yet understood, it is trivial. Rather, this is common in science. In other words, science never provides perfect explanations.
(b) The argument "We do not yet understand" needs to proceed with a mechanism in which the fact that "We know this much" is constantly recognized by society. Otherwise, everything can be turned into "unknown," which can easily be used as a justification for inaction.[12]

Science education does not teach that science cannot provide perfect explanations. Since it was feared that such facts would portray science as a "defective

---

[11] Tsuda (2004) links this with "reductionism to element." However, for the counteraction to work, it is only necessary to attack the flaws in explanation of a cause at any level and it does not have to be criticism of the type "causes have not been reduced to elements."

[12] *Kumamoto Nichinichi Shimbun* editorial from August 7 states the following before citing the aforementioned section. "They cannot take measures, given that they do not know the cause yet. They decided it was mercury and thus made improvements to the management of industrial wastewater, but if Minamata disease did not end up disappearing later, the irresponsibility of doctors and scientists will be the subject of endless criticism."

product," the "completeness" of science has instead been emphasized. However, while this model of science is (was) effective in countering pseudoscience, it is not appropriate for dealing with social issues related to science and technology. Rather, reverse aspects or side effects of emphasizing the completeness of science become all the more apparent.[13]

6.3.2 Criticisms over Repeated Backtracking

It is not hard to imagine how criticisms over Kumamoto University's repeated backtracking diminished people's trust in the research team's accuracy. It is even more so because at that point in time, organic mercury (methyl mercury) had yet to be determined as the cause of Minamata disease.

Repeatedly backtracking prompts, for instance, fickleness, defection, and evasion. Even in newspapers today, we see the word *backtracking* being used with clearly negative connotations: headlines in the *Asahi Shimbun*: "Passenger Anger Rises/Company Backtracks repeatedly over its Explanation/Ferry Drifting off the Shore of Akita"; "Defense Agency list/repeated backtracking over explanation of facts may affect Director's future"; "repeated backtracking over explanation/Yukijirushi Press Conference on mass food poisoning from low fat milk." Thus, criticism that Kumamoto University's group repeatedly backtracked became a foothold for dispelling suspicions for those who did not want the factory to be held responsible: "Organic mercury this time, but their theory may change again soon. No, I'm sure it will."

However, what was the real picture? Did the Kumamoto University research group repeatedly backtrack and revise their theories with no guiding principles, as suggested by *Mizu* and *Minamata Jiji Shinpo*? The answer is no.

Each researcher (chair) in the Kumamoto University research group concurrently considered a diverse range of heavy metals, and the metals that were believed to be particularly noteworthy changed during the course of research. It was not the case that the research group had some unanimous opinion and that opinion kept changing. If we examine their research as a whole, we observe that it was a process of gradually narrowing the range of possible metallic substances; in fact, they succeeded in doing so. By no means were they fickle in their judgments.

---

[13] Some may be of the opinion that how people understand science is a factor that cannot be ignored when compared with the "power" of the national government, prefectures, or corporations. However, no matter how tyrannical the power, they use people's "understanding/persuasion" effectively. Moreover, to obtain understanding/persuasion, they appeal to their commonly accepted ideas. In the case of the Minamata disease, they could "persuade" people by saying something along the lines of, "We are sorry about what is happening to these patients, but it is difficult to do anything until the cause has been completely worked out scientifically." Recent research on the public understanding of science from the perspective of rhetoric (Gross 1994, etc.) has provided many insights on these points.

Why was it, then, that the accusation of repeated backtracking was so effective or persuasive in public opinion?

One probable reason was that people were not informed that the group was gradually narrowing the possibilities. In other words, when the group changed the substance of focus, the public were not informed of their reasons. Any reports with such information are absent from newspaper articles. Neither is there any evidence that the scientists actively informed people or journalists of their reasons.[14]

Another related reason is that no one discussed the existence of alternative theories or possibilities. The researchers always spoke as if the situation were monolithic; for instance, when their attention focused on manganese, the group neglected to mention that other metals were also under consideration. Manganese was mentioned as if the group had already reached a conclusion. Thus, when thallium became the next suspect, it appeared as if the group's theory had suddenly changed.

These problems are connected with how scientists present their findings and how the media reports them. Confusion can be avoided when scientists correctly present their findings and the media correctly reports them. In early 2003, when the space shuttle Columbia disintegrated as it re-entered Earth's atmosphere, NASA repeatedly backtracked on its opinion of the disaster's cause. On February 7, 2003, the *Asahi Shimbun* newspaper ran the following headline: "NASA repeatedly backtracks over Opinion on Accident Cause/U.S. Shuttle Crash." Opinion kept seesawing as to whether the breaking off of heat-insulating material from an external fuel tank caused the accident. However, this did not cause loss of trust in NASA's investigations. NASA openly declared why the cause could or could not have been the insulating material breaking off. NASA also presented reasons to why it could not establish a single opinion (or why its opinion kept changing). The media accepted this and reported on the incident as follows: "From the outset, NASA has been clear that new facts are coming to light every day and that their opinions may change.... If one listens carefully to what Dittemore says, his use of such phrases as 'at the present time,' 'we cannot say for certain but...' and 'maybe...' stands out (*Asahi Shimbun*, February 7, 2003)".

So, what about the Minamata disease case? Simply put, it seems there were problems that cannot be reduced to how the group presented its results and how these were covered in the media. We can understand that the scientists were not inclined to report alternative theories to the world. For example, when the organic mercury theory surfaced in the summer of 1959, some researchers strongly supported the theory of thallium, but this was hidden from the outside world. Furthermore, opinions were divided among researchers. Supposing it was mercury, should they draw the conclusion of organic mercury and not inorganic mercury, or should they retain the possibility of inorganic mercury? Despite these concerns, the

---

[14] The articles of the researchers published in, for example, the *Journal of the Kumamoto Medical Society* contain their own individual reasons. However, the general public never read these, and they were printed and published much later.

Kumamoto University researchers unanimously declared that the cause was organic mercury (Food Sanitation Investigation Council Special Committee on Minamata Food Poisoning, September 8, 1959).

The research group, no doubt, acted in this way because they felt public admission of an uncertain conclusion would weaken their persuasive power. Moreover, we can surmise that behind that decision was the general public's perception that "Scientific endeavor should produce clear and unequivocal results." In other words, admitting ambiguities and uncertainty surrounding the theory would result in greatly reducing its persuasiveness.

In their day-to-day research, scientists are completely aware that science cannot provide perfect understanding. When speaking with colleagues at conferences, they use statements like, "We don't understand yet how things will develop from here on, but we do know this much." However, when speaking to a general audience, scientists tend to adopt statements easier for the general public[15] to accept—that express greater certainty—on the grounds that the former statement would be difficult for the public to accept.

However, such "internal affairs" becoming public can actually harm scientists' credibility. In the Minamata Disease Incident, the factory launched a prompt counterattack by stating, "Scholars are at loggerheads with their own group members even over the pathological findings"; this is why "it is hard to say that their findings are objective."[16]

The existence of "incompatibilities" between the general public's understanding of science and the handling of situations by scientists or a governmental administration often creates confusion. However, even when scientists or the administration handle the situation in a way that conforms to the general public's understanding of science, confusion still arises. Therefore, the general public's understanding should be corrected to something in line with the realities of science.[17]

6.3.3 Resonance Between Administrative Authorities and Science

This section focuses on what should be termed "resonance" between administrative authorities and science.

---

[15] Including people involved in administration.

[16] Four personnel from the factory attended the Food Sanitation Investigation Council Special Committee on Minamata Food Poisoning on September 8, 1959, implying the fact of conflicting opinions within Kumamoto University was leaked to the factory.

[17] The situation described here is similar to that described in Nakanishi (2004): "The dilemma is that without the notion of risk acceptance, only those risk assessments which depend of risk management are possible and can be published."

Administrative authorities need to have bases for implementing their policies. In many cases, science provides a convincing basis. However, when a scientific conclusion has not been obtained, authorities tend to hesitate taking concrete measures. Administrative authorities do not take action—in their words, cannot take action—if they are not certain or some points are left unexplained. In the Minamata Disease Incident, too, authorities did not take action on the grounds that they "did not know for certain."

To some degree, the authorities' response was evasive, given that the case involved legal responsibility and compensation. Besides that, they wanted to prioritize industrial protection. However, this alone does not account for their lack of response. In other words, even when such factors are involved, administrative authorities generally tend not to formulate measures on items they do not completely understand.

We see a good example of this in the response of Tadanori Ishiguro, a former government official (Military Medicine Headquarters). Before it was known that a vitamin B deficiency caused beriberi, the army discussed whether to change beriberi patients' meals from a rice diet to a wheat diet to reduce the number of patients. At this time, Ishiguro did not believe there was any particular conflict of interest in adopting a wheat diet. Despite this, he responded as follows:

> Rice has been our nation's staple food for thousands of years. If rice were harmful, why don't more Japanese people suffer from beriberi when only a minority of the total population do so? (After all, there are many other kinds of diseases with far more sufferers nationwide than beriberi.) Why do many men, but so few women, contract this disease? Why are many sufferers men in their 20s and 30s and so few in their 40s? Why are there especially large numbers of people around 20 years of age living in lodgings or group accommodation, such as students or soldiers? Why does the number of cases fluctuate wildly on an annual basis? To issue an order for something so uncertain—an order for the Japanese people to stop eating rice as their staple food and move to wheat, one [an order] at the suggestion of a health office at that—such an order declaring that this was for the prevention of beriberi would need to be made based on sound scientific bases. Rice would need to be subjected to microscopic analysis to see if it were harmful, along with sound and precise physiological experiments; only after approval from the scientific community could such an order be made (Ishiguro 1905).

In this way, administrative authorities search for scientific bases for policies related to questions of science—in other words, a definite conclusion. Moreover, if they fail to acquire a definite conclusion, they adopt a "waiting stance."

Meanwhile, science, by its nature, has a characteristic of indifference to the time required for solving a problem.

Science seeks to solve the mysteries of the natural world. Scientists are not satisfied by simply attaining a certain level (for example, satisfaction at making something, using things known thus far); rather, they seek an even deeper understanding, aiming to uncover new mysteries or problems. Time is no issue in solving these mysteries. So long as scientists know something is not a theoretical impossibility, they continue working on it, no matter how long it takes to reach a solution. In this sense, scientific research is, by its very nature, "timeless." The time required to find a solution is irrelevant. This sets it apart from engineering research or technological development. Awareness of the time required to solve a problem is a concept alien to scientific research.

Of course, this is not to say that individual scientists are unaware of the time required to solve a problem. For instance, in some cases, they are in competition with other researchers. To win this competition, researchers design procedures to produce results more quickly than others, perhaps with research themes that produce results at an early stage—chosen from the outset. In such cases, scientists are certainly aware of time. However, such cases also force scientists to rush their research and select themes according to social conditions; as such, they are not considered to align with the "essential nature" of science, wherein value is not awarded to the production of quick answers. As evidence of this, the kinds of problems or topics that result in quick answers are viewed as "mundane," with more "important" problems predicted to require more time to find an answer. Moreover, those researchers bold enough to tackle "difficult problems" are excellent, first-rate, and competent researchers. Therefore, in the absence of external factors, scientists are not aware of the time required to solve a problem.

It seems, then, that this essential nature of science and scientists actually creates problems in the real world. If scientific research is confined to the laboratory and not connected with external society, the research's timelessness is no problem. However, once science connects with social issues, timelessness becomes a major social problem. Timelessness means not caring about finding a solution "in time" and single-mindedly plowing on, regardless. This generates society's great expectations that a solution will be reached soon. The greater people's trust in science, the stronger these expectations—leading to predictions of a solution within a short time.[18]

Furthermore, when timelessness (the inability of science and scientists to rest while searching for elusive perfection) and the "waiting stance" of administrative authorities seeking definite scientific conclusions combine, they reach a stage of "resonance," an equilibrium wherein progress comes to a standstill. This is certainly what occurred in the Minamata Disease Incident.

The administrative authorities did not implement measures because "some things are still unexplained and scientifically uncertain." Thus, the scientists, burning with a sense of mission, made frantic efforts to solve the unexplained aspects. However, they needed more time to do so and to demonstrate the definite aspects. (Some problems have taken 40 years to solve.) The scientists became absorbed in their research with their pride at stake. Ironically, however, the greater their absorption, the more they became aware of their uncertainty—as if they had inadvertently responded to a provocation.

---

[18] Even today, the same kind of situation seems to be occurring in regard to iPS cell research. iPS cell researchers talk about a range of different things that will be possible with advances in research. However, they fail to mention how many years down the line these things will become possible. Moreover, by excluding time factors and talking about theoretical possibilities, they have generated a "dream for the near future" in society.

The author does not wish to criticize scientists for their devotion to finding causes. The intention is merely to suggest that when considering the governance of science and technology, greater attention should be paid to the fact that activities that seem natural to scientists can have unexpected effects in society, i.e., prolongation of administrative authorities' inaction through resonance.

Taking administrative authorities' waiting stance into consideration, the principle of clarifying everything currently understood and moving promptly to adopt a policy that is considered most appropriate at that point in time—given that predicting any date of complete understanding is difficult—will be essential.[19]

References

Gross, G. (1994). The roles of rhetoric in the public understanding of science. *Public Understanding of Science, 3*, 3–23.

Hosokawa, H. (1959). Hosokawa Hajime nohto (Hajime Hosokawa's notes). In Minamata-byo kenkyu-kai (Ed.), Minamata byo jiken schiryosyu:ge (Collected materials on the Minamata disease incident), (1996, p. 1579) Fukuoka: Ashi shobo.

Ishiguro, T. (1905). Rikugun eiseibu kyuhji-dan (Memories on the army health division), cited in Itakura, K. *Mohou no jidai: jyou* (The age of imitation: vol. 1). Tokyo: Kasetsusha, 1988, p. 372.

Kousei Kagaku Kenkyu Han (1957, March 30). Kumamoto ken Minamata chihou ni hassei shita kibyou ni tsuite (Regarding the mysterious illness which broke out in Minamata region, Kumamoto prefecture). Comprehensive research report funded through a Health Science Research Grant, Ministry of Health and Welfare. Also available in Minamata-byo kenkyu-kai (Ed.), *Minamata-byo jiken shiryo-syu* (Collected materials on the Minamata disease incident), (1996, pp. 833–845). Fukuoka: Ashi shobo.

Minamata-byo Kenkyu-kai (Minamata Disease Research Committee) (Ed.) (1996). *Minamata-byo jiken shiryo-syu*: jou-kan (Collected materials on the Minamata disease incident: vol. 1). Fukuoka: Ashi shobo.

Minamata Factory (1959). Yuhki Suigin Setsu no nattoku sh enai ten: youyaku (Reasons why we are not convinced by the Organic Mercury Theory: summary). In Minamata-byo Kenkyu-kai (Ed.), *Minamata-byo jiken shiryo-syu* (Collected materials on the Minamata disease incident), (1996). Fukuoka: Ashi shobo.

Nakanishi, J. (2004). Risuku hyouka to risuku manejimento no arikata: BSE no jirei kenkyu (The nature of risk assessments and risk management: the case of BSE). *Shiso*, 963 (July) 16–35.

NHK Special & NHK news crew (1995). Chisso Minamata koujou gijutsu-sha-tachi no kokuhaku (Confessions of engineers from the NJNFC Minamata factory). In NHK Special/NHK news crew (Ed.), *Sengo 50 nen, sono toki Nihon wa*: dai 3 kan (Japan 50 years after the war: vol. 3). Tokyo: NHK Publishing.

---

[19] Some may think that for scientists, this is as good as acknowledging the defeat of science. However, if the situation is likened to the differences in how clinicians and basic medical researchers deal with diseases, we see that it by no means acknowledges the defeat of science.

Nishimura, H., & Okamoto, T. (2001). *Minamata-byo no kagaku (The science of Minamata disease)*. Tokyo: Nihon Hyoronsha.

Sugiyama, S. (2005). Kagaku komyunikehshon (Science communication). In T. Nitta, et al. (Eds.), *Kagaku gijutsu rinri wo manabu hito no tameni (A handbook for those who learn ethics in science and technology)*. Kyoto: Sekai shisou sha.

Tomita, H., & Ui, J. (1969). *Minamata-byo: Minamata-byo kenkyu-kai siryo (Materials of the Minamata disease research committee*, Minamata-byo wo kokuhatsu suru kai (not for sale).

Tsuda, T. (2004). *Igaku-sha wa kougai jiken de nani wo shitekitanoka? (What have medical scientists been doing in pollution incidents?)*. Tokyo: Iwanami Shoten.

Ui, J. (1968). *Kougai no seiji-gaku: Minamata-byo wo otte (The politics of pollution: following traces of Minamata disease)*. Tokyo: Sanseido (Sanseido Shinsho).

Chapter 7
Itai-itai Disease: Lessons for the Way to Environmental Regeneration

Masanori Kaji

Abstract Itai-itai disease was first noticed in the Junzu River basin region in Toyama Prefecture in central Japan around the 1930s. However, it was not identified as a cadmium poisoning disease until the 1960s. A local physician, with cooperation from outside experts, confirmed that the disease was caused by pollution from the Kamioka mine of the Mitsui Mining and Smelting Co. Ltd., located in the upstream region of the river. In the mid-1960s, the victims of Itai-itai disease filed a suit against the company and won their case in 1972. The victims received compensation and signed a pollution control agreement with the company. The case of Itai-itai disease is a rare example of successful pollution control in Japan, because the ensuing 40-year annual inspections, based on the pollution control agreement, show that cadmium concentrations in the river have been reduced to natural levels. By analyzing the roles of various experts involved, this case study has contributed substantially to the understanding of the nature of expertise and the significance of public participation in the resolution of environmental problems. The author suggests some lessons learned from the Itai-itai case are applicable to the environmental regeneration of areas degraded by the recent Fukushima nuclear disaster.

7.1 Introduction

During the 1960s in Japan, people were beginning to realize that science and technology are not always beneficial for social progress, but instead can sometimes cause problems. During that period, Japan was enjoying very high economic growth, but at the same time was suffering from different forms of pollution, such as

---

This is a revised and expanded paper based on earlier versions in Japanese by Kaji (2005, 2009) and, in English, Kaji (2012).

M. Kaji (✉)
Tokyo Institute of Technology, Tokyo, Japan
e-mail: kaji.m.aa@m.titech.ac.jp

© Springer International Publishing Switzerland 2015
Y. Fujigaki (ed.), *Lessons From Fukushima*,
DOI 10.1007/978-3-319-15353-7_7

air, water, noise, and mining pollution. Japan was even referred to as "one of the advanced nations of the world in pollution."[1] Consequently, pollution-related diseases became a matter of public concern. The affected residents and victims began to sue polluters in an attempt to have their complaints made public and to receive compensation for damages. This included the so-called "Four major pollution-related lawsuits" in the 1970s: Itai-itai disease, Minamata disease, the Second Minamata disease in Niigata Prefecturee, and Yokka-ichi Pollution.[2] All the victims filed suits against the companies that were responsible for their injury and they won their cases.

This chapter analyzes Itai-itai disease as an example of the various attempts that have been made to find causes and to solve environmental problems, because the Itai-itai case was the only one out of the four major lawsuits where pollution was successfully controlled almost completely. The victims succeeded due to a multiple year movement by patient groups in cooperation with specialists and concerned citizens. This rare example of successful pollution control in Japan can provide good lessons for recovery from the environmental catastrophe caused by the Fukushima Daiichi nuclear power plant accident.

Itai-itai disease was a pollution-related disease caused by a very severe type of cadmium poisoning arising from the pollution of rice fields. The liquid wastes of the Kamioka mine of the Mitsui Mining and Smelting Co., Ltd. (Mitsui Kinzoku) were eventually incriminated as the source of the cadmium. Kamioka was one of the richest zinc mines in 20th century Japan and the cadmium, which occurs as a minor component in most zinc ores, was a byproduct of zinc production (Daintith 2008, p. 91).[3] The numbers of patients suffering from this disease between 1910 and 2008 were estimated at approximately 400 (Matsunami 2010, pp. 44–48, 537).[4]

---

[1] In his Special Message to the Congress on Environmental Quality, February 10, 1970, U. S. President Richard Nixon said vaguely that "by ignoring environmental costs we have given an economic advantage to the careless polluter over his more conscientious rival" (Nixon 1971, p. 96). Michio Hashimoto, the head of the pollution department at the Ministry of Health and Welfare of the Japanese government then, thought that Nixon implicitly criticized Japan as such an unfair polluter (Hashimoto 1988, p. 157).

[2] There is a section "Yondai Kogai Saiban no Kyokun (in Japanese, Lessons of the four major pollution-related lawsuits)" in the White Paper on the Environment in Japan for 1973 (Kankyo-cho 1973). The section describes each lawsuit and emphasizes the importance of anti-pollution measures by private companies in industry in Japan.

[3] Cadmium is located just below zinc in the Periodic Table and its properties are similar to those of zinc. Most cadmium in nature occurs as atomic substitution for zinc in zinc minerals and it is produced as an associate product when zinc ores are reduced. http://www.mii.org/Minerals/photocad.html (accessed July 30, 2014).

[4] The book by Matsunami (2010) is the most comprehensive and readable book that analyzes every aspect of Itai-itai disease and related cadmium poising in Japan. The book was first published in 2002 as *Itai-itai byo no kioku* (The Memory of Itai-itai disease) with around 230 pages. However, the book was revised and expanded successively for a short period and became a more than 600-page thick cyclopedic book. Matsunami lists 195 officially designated victims of Itai-itai disease between 1967 and 2008, and says that around 200 victims must have died before 1967, when the official system of certification of victims was initiated.

Notably, the majority were women.[5] The reasons for this are most likely because women have lower body weights and less dense bone mass in general and their physiology specifically enhances the uptake of cadmium and therefore its damage (Matsunami 2010, pp. 451–461).

The case of Itai-itai disease will shed light on problems with expertise and public participation, especially regarding the role of experts and the significance of public participation in resolving techno-environmental problems.[6]

7.2 The Kamioka Mine and Mining Pollution

The Toyama basin, located in central Honshu (Japan's main island), has been one of the richest rice producing areas since the Middle Ages, and the Jinzu River is one of the two main rivers running through this basin (the other is the Sho River) (see Fig. 7.1). The Kamioka mine, which is located in the upstream region of the Jinzu River, was first developed in the 17th century as silver, copper, and lead mine (see Fig. 7.2).

After the Meiji Restoration in 1868, which marked the beginning of the Westernization of Japan, the Kamioka mine was purchased by the Mitsui Company,[7] one of the largest private capital groups. By 1889, Mitsui had purchased all the pits and the operation rights of the Kamioka mine. Beginning in 1905, Mitsui started to mine zinc in Kamioka. At first, Japan exported zinc ore and imported zinc metal, because the country had no smelting facilities. Zinc ore was initially separated based on density differences, but from 1909, zinc minerals began to be separated using surfactants and wetting agents (froth flotation). In 1914, in Miike, Fukuoka Prefecture, Mitsui introduced zinc ore smelting processes to produce zinc metal, the first in Japan.[8] In the 1920s, after World War I,[9] the Kamioka mine became one of the largest zinc producers in Japan.[10] Newly introduced, improved froth flotation

---

[5] There were only three male victims (1.5 %) out of 195 victims officially designated between 1967 and 2008 (Matsunami 2010, p. 453).

[6] For a good introduction to the recent literature on expertise and public participation, see, for example, Evans and Collins (2008) and Sismondo (2010, pp. 180–188).

[7] Mitsui Company was founded by the Mitsui family, a merchant family in the 17th century. After the Meiji Restoration, the family established Mitsui Company in 1872 and in 1876 the Company was reorganized into Mitsui Bank, Japan's first private bank and Mitsui Bussan, a trading company. These two companies together with Mitsui Mining, established in 1888, became the core enterprises of one of the largest corporate conglomerates (*zaibatsu*) in modern Japan.

[8] Smelters in Kamioka started to operate only in 1943 (Matsunami 2010, p. 88).

[9] Japan took advantage of the absence of European powers in Asia, due to the War, to expand its influence in Asia and in the Pacific. It enjoyed an economic boom and unprecedented prosperity, and the years between 1910 and the 1920s marked a turning point for the Japanese economy.

[10] Mitsui produced about 48 % of Japan's zinc metal after 1925 (Matsunami 2010, p. 88). The Kamioka mine produced 57 % of all zinc ore mined in Japan in the first six months of 1950 (Yamada 1951, p. 46).

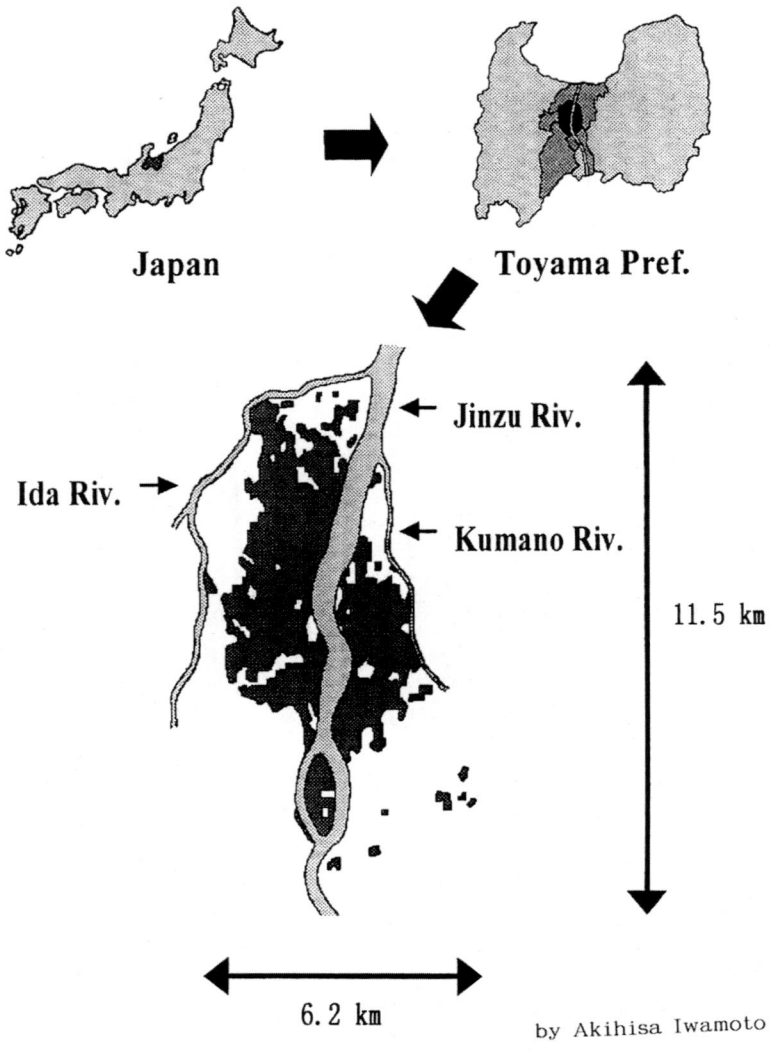

Fig. 7.1 A map of the polluted area in the Jinzu river basin of the Toyama plain (Iwamoto 1999, p. 180). With permission from the author's group

processes in the 1920s helped to increase zinc production, but some of the fine powdered mineral particles produced in the frothing process escaped and floated down the river. These fine particles were then easily oxidized into ions, which were, in turn, readily absorbed by plants and humans (Kurachi et al. 1979, p. 81). The waste from the Kamioka mine increased sharply during the 1930s, when the war with China created a significant demand for zinc production. Farmers and peasants

7 Itai-itai Disease: Lessons for the Way …

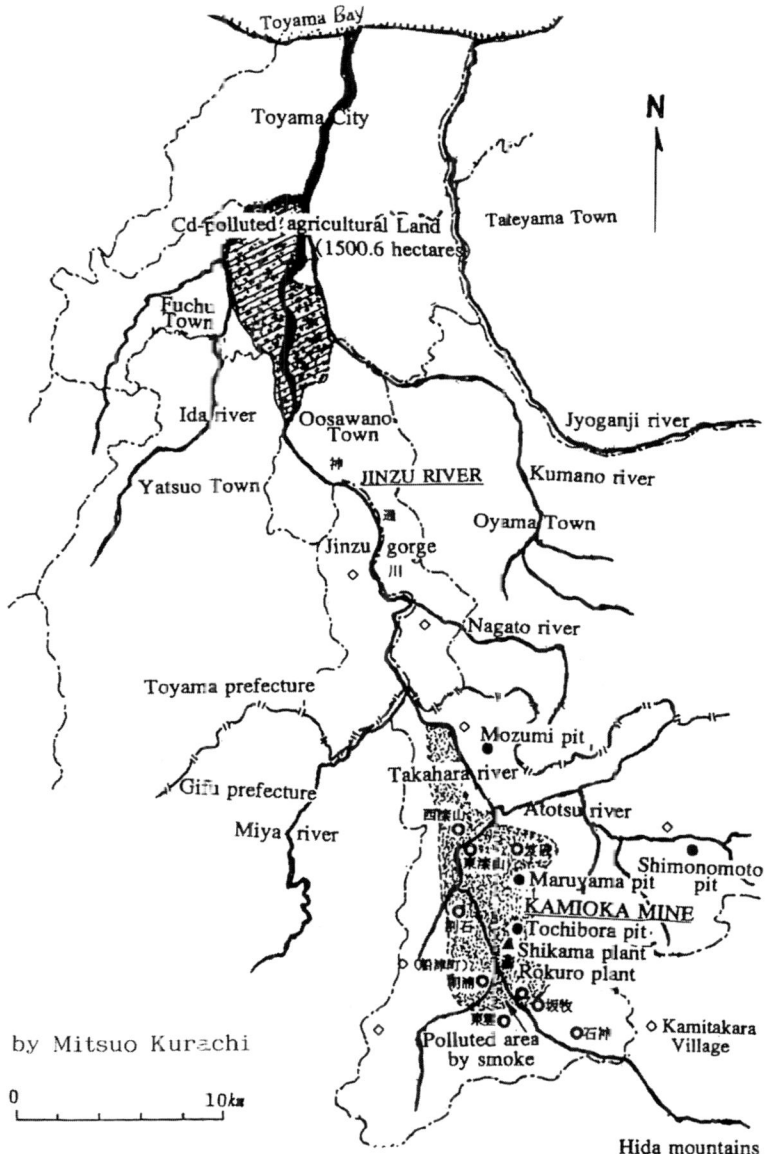

Fig. 7.2 The Jinzu River Basin and the location of the Kamioka mine (Kurachi 1999, p. 150 with slight modification). With permission from the author

who used the water of the Jinzu River for agriculture and fishermen who fished those same waters, noticed decreases in crop yields and fish catches and established an association in 1932 to protest mining pollution. During World War II, unskilled workers replaced skilled miners, who had been drafted. The result was an increase in yield loss, along with an increased volume of unnecessary waste, which was dumped into the Jinzu River. Agriculture damage in terms of rice production and the damage to the fishery industry noticeably increased.

Cadmium-containing ores are rare in nature, with greenocite (CdS) being the only major cadmium mineral. This mineral is almost always associated with sphalerite (ZnS), the main zinc ore in Kamioka. For this reason, cadmium was a common by-product of zinc ore mining. During World War II, cadmium had no industrial value,[11] so Mitsui simply discarded it as waste into the Jinzu River. Mitsui started to extract cadmium as part of the zinc production process in the Kamioka mine only after 1948.

7.3 The Discovery of Itai-itai Disease and Its Causes

Itai-itai disease was named and identified in 1950s, but the Minister of Health and Welfare estimated in 1968 that the first patients of the disease appeared as early as the 1910s in the area of the Jinzu River basin in Toyama Prefecture.[12] Shigejiro Hagino, a local physician, was one of the first to notice this unfamiliar disease, characterized by severe pain in all parts of the body around 1935, and correctly predicted mining pollution as its cause. The Hagino family was one of the prestigious and influential local landholders and, for generations, the family members had been physicians (Hatta 1983, p. 22). Noboru Hagino (1915–1990), Shigejiro's son, served in the army as a doctor soon after graduating in 1940 from Kanazawa Medical College (now the Medical School of the Kanazawa University), a national medical college in the city of Kanazawa. He returned to his home village in March 1946 from China, where he had served as an army doctor. He inherited a local clinic from his father, who had passed away in 1943, before his return.

Soon after World War II, Itai-itai disease, initially thought to be a rheumatic disease with no connection with mining waste problem, was reported in a medical journal by researchers at Kanazawa Medical College (Matsunami 2010, pp. 9–11). The 44 patients studied had an average age of 57.7 years and were mostly women. A more thorough medical study on the disease was conducted in 1955 by a group of doctors from a rheumatology institute in Tokyo. Noboru Hagino participated in this

---

[11] The first cadmium production for industrial use in Japan began in 1929. Cadmium was first used for pigments and as a component in low melting alloys. Today, cadmium is used mainly in batteries (especially Ni-Cd batteries) (Matsunami 2010, pp. 126–128).

[12] The Ministry provided this estimation in the announcement by the Ministry of Health and Welfare on May 8th 1968 (see the Sect. 7.4 of this chapter). Matsunami also confirmed this estimate (Matsunami 2010, pp. 40–48).

study as a local doctor, unaware of his father's earlier observations and not anticipating any connection with wastes from the mine (Matsunami 2010, pp. 11–14). The *Toyama Shinbun*, a local newspaper, gave the syndrome its name of Itai-itai disease,[13] since it was a very painful disease and patients, mostly women over 35 years old, often cried out in pain "Itai, itai" (meaning "It hurts, it hurts"). The study concluded that this disease was not rheumatic, but a kind of osteomalacia (Matsunami 2010, pp. 13–14).

Nobuo Hagino, like other physicians, first thought that overwork and malnutrition were the cause of disease, but soon realized that all the patients lived and worked near the midstream region of the Jinzu River, in an area irrigated by river water. Therefore, Hagino began to consider mining pollution from the Kamioka mine, located upstream, as a potential cause. In 1957, Hagino reported to an academic society that the cause of the disease was heavy metals, especially zinc, based on experimental work on the cause of diabetes by professor Kozo Okamoto (1908–1993) of Kyoto University in 1952 (Hagino 1968, p. 53). Professor Okamoto had proved that zinc chelating agent such as oxine (8-hydroxyquinoline) caused diabetes; however, Hagino thought that Okamoto had proved that zinc caused diabetes. Since most of Itai-itai disease sufferers show proteinuria (presence of an excess of proteins in the urine) or glycosuria (excretion of glucose into the urine), Hagino guessed that zinc caused Itai-itai disease. This was a wrong guess, because Okamoto had proved that oxine caused diabetes, not zinc. Oxine prevented the functioning of zinc bound to insulin, the hormone that plays an important role in sugar metabolism. Naturally, his report was heavily criticized by other researchers. Hagino's guess that mining pollution was a possible cause of the illness was right, but he could not show sufficient evidence to persuade researchers who believed that the disease was caused by malnutrition, overwork, and lack of sunshine during the winter.

In 1948, two organizations with the same name, the Jinzu Mining Pollution Prevention Council (JMPPC), were established: one led by heads of local administrations and the other led by heads of agriculture cooperatives. Japan has a history of conflict between groups of farmers and fishermen and large mining companies from early 20th century various parts of Japan.[14] It was not unusual for companies to pay some compensation money to the complaining groups in damaged area. Kamioka mine paid some money to complaining groups every year from 1949 to 1954 after negotiation (Matsunami 2010, pp. 153–154). Meanwhile, the two Mining Pollution Prevention Councils merged in 1951 and the new organization

---

[13] *Toyama Shinbun*, August 4th 1955. See also the photo reproduction of the article in Matsunami's book (Matsunami 2010, p. 10).

[14] Shozo Tanaka (1841–1913), a statesman and a member of the House of Representative of Japan, was famous as a leader of protest movement against the Ashio Copper Mine in Tochigi Prefecture that caused the agricultural and fishery damage in the area downstream. On Shozo Tanaka, see, for example, Yui (1984), Dehn (1995).

demanded compensation from the Kamioka mine. In 1955, the three parties (the JMPPC, Toyama Prefecture, and the Kamioka mine) reached an agreement regarding the amount of compensation, which was to be revised every five years.[15] In 1960, the next compensation agreement was concluded with less money being issued. The Fuchu-machi branch of the JMPPC decided to commission Kin-ichi Yoshioka to carry out research on the damage caused by mining pollution. The aim was to find stronger evidence against the Kamioka mine, which had tried to reduce compensation to the victims in every possible way. Yoshioka was a well-known expert on agricultural flood damage, and in September 1958, he happened to visit the Jinzu River to conduct research on flood damage as well as on agricultural damage caused by mining pollution. In August 1960, the Fuchu-machi branch of the JMPPC commissioned him to do a full survey of the agricultural damage.

Yoshioka decided to extend his study of agricultural damage to include other forms of possible biological damage, including Itai-itai disease, which at that time was still a local disease of unknown cause. He used an epidemiological study to plot the locations of known Itai-itai disease patients on a map, with the help of Hagino. He clearly demonstrated that the disease appeared in those places that were irrigated by the Jinzu River. His analysis of the metal content of water, soil, vegetables, animals, and the organs of the patients revealed an unusual abundance of cadmium. He noticed a correlation between the annual increase in numbers of patients and the increased production of zinc ore at the Kamioka mine. He also found a similar case of chronic cadmium poisoning in a French medical journal (Nicaud et al. 1942). In June 1961, in his report to the head of the Fuchu-machi branch of the Council, Yoshioka concluded that the cause of Itai-itai disease was cadmium discarded by the Kamioka mine (Yoshioka 1961).[16] At the same time, he reported these results, together with Hagino, at a scientific meeting of a medical society (Matsunami 2010, pp. 28–29).

Professor Jun Kobayashi (1909–2001),[17] an analytical chemist at Okayama University, helped with the detection of cadmium in the samples sent by Yoshioka and Hagino. He developed specific heavy metal analysis techniques to advance the study of Itai-itai disease. These three experts, with their different backgrounds, played a decisive role in identifying the cause of Itai-itai disease.

---

[15] The compensation money was only for the agricultural and fishery damages, not for the pollution-related disease. The connection between the Itai-itai disease and mining pollution was beginning to be noticed by the mid-1950s and was confirmed in the early 1960s.

[16] Yoshioka later published his conclusion in a local medical journal (Yoshioka 1964).

[17] Jun Kobayashi was born in Kurashiki, Okayama Prefecture. He graduated from the Agricultural Department of Tokyo Imperial University. After working at agricultural experimental stations of the Ministry of Agriculture and Forestry, he worked as research associate at Ohara Agricultural Institute. After World War II, he became a faculty member of Okayama University, a national university in Okayama Prefecture. He retired in 1975.

7.4 The Central and Local Administrations on Itai-itai Disease

The central government and the local authorities were not very cooperative in their responses to the victims of Itai-itai disease. At the end of 1961, Toyama Prefecture set up a Special Committee for Prevention of Local Diseases, the first of this kind on Itai-itai disease by a local authority. The committee made a list of patients who were then surveyed using questionnaires that reinforced the malnutrition theory of the disease and defused the idea that mining pollution was the cause of the disease. The Toyama Prefecture administration, in the late 1950s, was trying to promote industrialization in the middle of the Junzu River Basin of the Toyama Plain. They feared that the emergence of a pollution-related disease like the Itai-itai would dissuade enterprises from coming to the region (Matsunami 2010, pp. 30–32).

In 1963, government-funded research groups were established under the auspices of the Ministry of Health and Welfare and the Ministry of Education to look into the cause of the disease. In January 1967, the joint research group of both Ministries admitted that cadmium was one of the causes of the disease, but claimed that other causes might also exist (Matsunami 2010, pp. 32–35). Victims were disappointed by this vague conclusion by the governmental research groups and in March 1968 they decided to file a case against Mitsui and the government.

During this period, the Ministry of Health and Welfare was beginning to realize the need for a new approach to industrialization and urbanization in Japan. When the pollution department was created in the Ministry in April 1964, Michio Hashimoto (1914–2008) was appointed as its first head of the department. Hashimoto belonged to the first generation that was aware of new approaches in public health and he played a key role in changing the Ministry's attitude.[18] The Ministry commissioned further study with the Japan Public Health Association (JPHA), the oldest non-governmental public health organization in Japan,[19] to investigate the aspects that the joint research group of the two ministries did not cover. In the spring of 1967, a group of researchers was organized by the JPHA, headed by Itsuzo Shigematsu, chief of epidemiological department of the Institute of Public

---

[18] According to Hashimoto's autobiographical reminiscences (Hashimoto 1988, pp. 19–66), he was born in Osaka in 1924. He graduated in 1948 from the Medical School of Osaka University and worked as doctor in the Public Health Center of the Osaka Prefecture. He was first sent to study at the National Institute of Public Health of the Ministry of Health and Welfare in Tokyo for a year and, in 1954, he was sent to the United States for one year to study public hygiene (a Master's course) at the Harvard School of Public Health. After returning to Japan, he was recruited in 1957 by the Department of Public Health of the Ministry of Health and Welfare. In 1972, he moved to the Environment Agency, established in the previous year. After an early retirement in 1978 he became professor of Public Health at Tsukuba University, which was a newly established national university.

[19] The Association was first established as *Dai-Nihon Shiritsu Eisei-kai* (The Great Japan Private Public Health Society) in 1883.

Health, who later played a negative role (see Sect. 7.7). The group included professors of public health, analytical chemistry, and geology at Kanazawa University, researchers from Toyama, and Jun Kobayshi from Okayama (see the previous section). They showed a correlation between the appearance of patients with Itai-itai disease and the use of the Jinzu River water, as well as the degree of cadmium pollution in the soil (Matsunami 2010, pp. 75–77).

Hashimoto insisted that the government should act based on available scientific evidence, but should not wait for conclusive scientific proof, to avoid an untimely decision (Hashimoto 1988, p. 136).[20] His opinion was accepted by the ministry. Based on the results of various investigations, especially those of the group of researchers organized by the JPHA (Matsunami 2011, p. 79), on 8 May 1968, the Minister at the time, Sunao Sonoda (1913–1984), announced that the disease was caused by cadmium poisoning and the only possible source of the cadmium was the effluent coming from the Kamioka mine.

7.5 Resident Movement and the Itai-itai Disease Trials

Mining damage to agriculture occurred much earlier than damage to human health in the form of Itai-itai disease. As early as 1890, sulfur dioxide gas from roasting furnaces (used to reduce sulfur content and heavy metal powder dust) and acid rain resulted from the gas had harmful effects on plants, aquatic animals and infrastructure in the area surrounding the Kamioka mine. From 1896, agricultural and fishery damage, such as reduced crops and catches, dead fish became a problem in the midstream and downstream areas of the Jinzu River. Around 1932, the Jinzu River Mining Pollution Prevention Union was established and the Union asked the prefectural administration to analyze the river water and demanded that Mitsui Mining & Smelting Co., Ltd build preventive facilities (Kurachi et al. 1979, pp. 109–111).

The Jinzu Mining Pollution Prevention Council (JMPPC) was established soon after World War II to counter the pollution in the villages in the downstream area of the Jinzu River. The Council's focus was on agricultural damage; it was not concerned with pollution effects on humans (Fujikawa 2005, pp. 108–110).

After Hagino and Yoshioka reported at a 1961 academic meeting that the cause of Itai-itai disease, based on well-founded research, was cadmium discarded by the Kamioka mine, a number of research groups were established by the Prefecture as well as the central government. The locals, including Hagino, expected a fair

---

[20] Hashimoto referred to the case of Minamata disease as this type of untimely decision (Hashimoto 1988, p. 145). This was one of earliest announcements in Japan by the government officials based on what later became known as precautionary principle. Even earlier, Mitsuo Taketani (1911–2000), a theoretical physicist and philosopher, stated that one should not do nuclear weapon tests without the proof of their safety, while military leaders overseas claimed that they could because of lack of scientific proof of their danger (Taketani 1957, p. 36). One can consider Taketani's idea as a precursor of the principle.

solution by the authorities.[21] However, the joint research groups of the Ministry of Health and Welfare and the Ministry of Education presented inconclusive evidence and mentioned the possibility of cadmium as the cause of the disease only in passing. This disappointed the patients, who expected conclusive results from governmental research. Young local leaders in the polluted areas, together with Hagino, realized that they had to act on their own behalf in order to solve the Itai-itai disease. In November 1966, the Itai-itai disease Residents' Association was established in the Kumano District of Fuchu-machi, Toyama Prefecture, which was the area with the highest pollution level and the highest concentration of Itai-itai patients. The aims of the Residents' Association were to obtain relief for patients, a free water supply, remediation of polluted soil, and clarification of the company's responsibility for the disease (Matsunami 2010, pp. 162–171). The members of the Association tried to negotiate directly with the Kamioka mine several times in the summer of 1967 on the issue of compensation for the dumping of the polluted effluent, but the company denied the dumping charge, stating that no scientific proof existed to show a causal relationship between the company's activity and the disease. A representative of the Kamioka mine even said that the company would willingly compensate every incident of damage if official institutions could verify any responsibility of the "world-renowned large" company, Mitsui, for the disease (Itai-itai byo sosho bengo-dan 1972, p. 585; Matsunami 2010, pp. 171–172).

In the 1960s, several acute cases of pollution-related disease appeared in Japan and attracted nationwide attention. Of these, Minamata disease, a neurological syndrome caused by severe mercury poisoning, was the most famous. This disease was first discovered in 1956 in Minamata City, in Kumamoto Prefecture in Southern Japan. It was caused by methyl mercury found in the industrial waste from the Chisso Corporation's (the New Japan Nitrogenous Fertilizer Company) chemical factory in Minamata. In June 1965, patients with a similar mercury poisoning disease were found in Niigata Prefecture, a neighboring Prefecture to Toyama Prefecture and this was referred to as the second Minamata disease. Its victims and residents organized a victims' association in August of the same year, as they were unhappy with the sluggish action of the government. The association filed a suit in June 1967 against Showa Denko, the company responsible for the mercury contamination, as the first of four major pollution-related lawsuits.[22] Some members of the Itai-itai disease Residents' Association participated in the site inspection for the lawsuit in October 1967 and identified the need for a lawsuit.

At the end of 1967 the Itai-itai disease Residents' Association in the Kumano District of Fuchu-machi, founded a year earlier, held a meeting in Fuchu-machi to discuss a lawsuit. The meeting was attended by two young lawyers from Toyama

---

[21] Hagino was much criticized and was under pressure not to work on Itai-itai disease. He even almost gave up studying the disease and he traveled to the United States and Europe to escape the unsettled situation during 1963 and 1964 (Matsunami 2010, pp. 158–159).

[22] See the introduction to this paper. The Yokkaichi pollution disease patients followed in September 1967. Finally, after initiation of the Itai-itai disease case, Minamata disease victims filed a suit again Chisso Cooperation, the source of mercury pollution, in June 1969.

Prefecture, Tatsuru Shimabayashi (1933–) and Jun-ichi Matsunami (1930–).[23] After vacillating for some time, the Residents' Association, with the help these young lawyers, decided to file a lawsuit against Mitsui in January 1968.

Shimabayashi, a member of the Japan Young Lawyers Association, a large liberal association of lawyers in Japan, asked for the help of other lawyers in the organization. Matsunami succeeded in gaining the support of local lawyers, including those with leftist or liberal tenets as well as those with conservative beliefs. Among these local lawyers was Kinosuke Shoriki (1904–1980), a respected and influential lawyer whose uncle, Matsutaro Shoriki (1885–1969), was a conservative politician and the president of *Yomiuri Shinbun*, a conservative national newspaper with the largest circulation in Japan. This connection would prove helpful in winning the lawsuit, as it brought on board conservative and respected persons like Kinosuke Shoriki to counterbalance the other members show shared largely antigovernment and leftist activist leanings.

On March 9, 1968, 14 victims as well as 14 family members of deceased victims filed a suit against Mitsui demanding 61 million yen for pain and suffering. This was the first group of representatives of patients or the bereaved from polluted towns. If this court fight went well, other victims would follow suit. In all, 236 lawyers, mostly young and inexperienced, but passionate and hardworking, were involved in the suit. Twenty worked on the suit full-time and attended every session of the court; Shoriki became the head of this group of lawyers. One of these lawyers, Chuko Kondo (1932–2013), moved to Toyama from Tokyo and set up his permanent office in front of Toyama station, which became a center for the court fight. Some lawyers, including Shimabayashi, followed him to Toyama. These moves helped in gaining local support.

The suit was one of the largest pollution lawsuits in the history of Japan. Mitsui, one of largest monopoly capital groups in Japan, employed the most prestigious lawyers. They even at one point tried to hire Kinosuke Shoriki to work for them. When Shoriki declined, Mitsui requested that he should not engage further in the suit (Matsunami 2010, pp. 213–214).

After 36 pleadings and four on-site verifications of the mine, the plaintiffs of the Itai-itai disease won their first case in June 1971. In the lawsuit, four experts (three university professors and Dr. Hagino) attested to the facts in favor of the plaintiffs. Noboru Hagino and Jun Kobayashi, two of the main discoverers of the cause of the disease, were joined by Arinobu Ishizaki, a professor of the school of medicine at Kanazawa University, and Saburo Fukai, a geologist and a professor of the Toyama University. Mitsui's only expert testimony was provided by the director of the Kamioka mine hospital.

---

[23] Matsunami is the author of a comprehensive book on the Itai-itai disease (Matsunami 2010; see also Note 3). Both Matsunami and Shimabayshi were from Toyama Prefecture and were still in their thirties; each had only one or two years' experience as lawyers at the time (Matsunami 2010, pp. 185–188).

The court decided—based on the epidemiological evidence—that a causal relationship existed between the disease and the liquid wastes of the Kamioka mine.[24] This was the first occasion of a judicial decision based on epidemiological evidence in Japan. In March 1970, one year before the court's verdict, the Supreme Court in Japan had gathered judges who had engaged in pollution-related lawsuits and they had studied the causal relationships in those lawsuits (Matsunami 2010, p. 257). They had admitted that proof based on epidemiological evidence was adequate in principle and if an offending company was unable to disprove the plaintiff's proof, a court should decide in favor of the plaintiff.[25]

Mitsui appealed to a higher court, with Jugoro Takeuchi (1922–1998), a professor at the School of Medicine at Kanazawa University, testifying for them. Takeuchi had once explained the mechanism of the function of cadmium in causing Itai-itai disease. He now denied his own theory and claimed that vitamin D deficiency was the cause of the disease. His earlier theory was crucial for the plaintiffs' victory in the suit, so if he could persuade the judges that he had been wrong, then that would deal a fatal blow to the victims' case. However Jun-ichi Matsunami, one of two young lawyers who played a key role in starting the lawsuit, cross-examined Takeuchi and was able to disprove Takeuchi's new argument (Matsunami 1998, pp. 55–131).[26]

We do not know the reason why Takeuchi denied his own theory in favor of Mitsui. One possible explanation is that Mitsui applied pressure and inducement. Takeuchi's vitamin D deficiency theory did not change the court's decision, but it played a definite role in the subsequent rollback by the mining industry, as discussed later. Takeuchi was one of many experts whose "expertise" was used to maintain the status quo of the ruling system and who were rewarded with high positions in return.[27]

The Kanazawa branch of the Nagoya High Court did not admit any further witnesses that Mitsui proposed. After only 12 pleadings, it dismissed the appeal on August 9, 1972, and the case ended in favor of the plaintiffs. Mitsui did not appeal further. The court's final decision was that Itai-itai disease was caused by cadmium discharged by the Kamioka mine and that Mitsui was responsible for compensating

---

[24] In fact, even though the Jinzu River flows through the Toyama basin, the only area ridden with the disease relied on river water for agricultural irrigation as well as for human consumption.

[25] This matter was discussed even in Diet (Japanese parliament) in 1970. See the 63rd Diet minutes of the Industrial pollution prevention special committee on April 1 (http://kokkai.ndl.go.jp/SENTAKU/syugiin/063/0620/06304010620006c.html) and the 64th Diet minutes of the legal committee on December 8th (http://kokkai.ndl.go.jp/SENTAKU/syugiin/064/0080/06412030080004c.html).

[26] Matsunami played an important role in various pollution-related lawsuits, including the SMON (subacute myelo-optico-neuropathy, a drug-induced disease) lawsuit and the Minamata disease (methyl mercury poisoning) lawsuit, until his retirement in 2001.

[27] The Itai-itai disease suit did not seem to cause any damage to Takeuchi's career. He was born in Chiba Prefecture and graduated from the medical school of Tokyo Imperial University in 1944. In 1974, he became a professor at the Tokyo Medical and Dental University, a prestigious national medical university in Tokyo, and he later became the Director of the University's hospital. He was even elected president of the Japanese Society of Internal Medicine, one of the oldest and most prestigious medical societies.

the victims for all damage, as well as for remediation of the contaminated soil (Matsunami 2010, pp. 299–302).

By the morning of August 10th, the day after the court's decision, negotiations between Itai-itai disease victims and the company had started at Mitsui's headquarters in Tokyo. After an eleven-hour confrontation, the company admitted that the disease was caused by cadmium and other heavy metals discharged by the Kamioka mine, and thereafter pledged to refrain from disputing the court's judgment. Mitsui also admitted its responsibility in other Itai-itai lawsuits filed by other groups of patients.[28] The company agreed to pay all the plaintiffs of the litigating groups and also to compensate all the sufferers of the disease, including those who were under medical observation and suspected of having the same disease. Mitsui also agreed to compensate for all the agricultural damage and to restore the polluted soil (Matsunami 2010, p. 303).

One of the most important victories was the establishment of the Pollution Control Agreement (Isono 1999), concluded between Mitsui and the victims' group. Under the terms of the Agreement, the victims' group, the Itai-itai disease Residents' Association, had the right to enter and inspect the mines and factories, together with experts, at the company's expense, at any time and whenever the Association considered it necessary. The company was obliged to release any data on pollution at the request of the Association. The company was also obliged to do its best to fulfill the requirements of the Residents' Association to improve its facilities in order to prevent further pollution.

Soon after its organization in 1966, the Itai-itai disease Residents' Association had demanded inspections of the Kamioka mine, but its representatives had always been sent away at the main gate of the mine (Matsunami 2010, pp. 168, 171–172). The first inspection was not conducted until November 1968, and only when ordered by the court. Even then, the association representatives and their experts had been unable to complete their inspection of the mine due to the uncooperative attitude of the company (Matsunami 2010, pp. 228–233). Therefore, the right to enter and inspect the factory was the one of the most important victories that resulted from the lawsuit.

7.6 The Role of Experts in the Rollback by the Mining Industry

Around 1975, the mining industry began to take action to counteract citizen movements against the industry. The backlash was initiated by an article, written by Takaya Kodama (1937–1975), a rising young journalist, who questioned whether Itai-itai disease was a pollution-related disease. Kodama was also critical of Noboru Hagino, one of the main discoverers of the cause of the disease (Kodama 1975). He

---

[28] There were six other lawsuits.

criticized the cadmium theory, basing his arguments on Takeuchi's vitamin D deficiency theory, even though that theory had been refuted as a cause in the Itai-itai disease suit in the High Court. Kodama even claimed that Itai-itai disease was a kind of iatrogenic disease, caused by excess administration of vitamin D by physicians.

Kodama himself died of cancer in May of the same year, at the age of 38.[29] However, the Japan Mining Industry Association distributed his article in many places and even translated it into English for overseas readers (Matsunami 2010, p. 315, footnote 2). Some parliament members from the ruling Liberal Democratic Party began to question the validity of the Ministry of Health and Welfare's announcement on the cause of Itai-itai disease.

The Environmental Agency, established in 1971 as a government agency, started a research program on Itai-itai disease and cadmium poisoning in 1974. After questioning on the credibility of the cadmium theory of Itai-itai disease by Zentaro Kosaka (1912–2000), a prominent conservative politician, in the House of Representatives, the Lower House of Japan, the Agency reorganized the research group, with one of its aims being the reexamination of the cause of Itai-itai disease (Matsunami 2010, pp. 315–316). The majority of researchers in the group doubted that cadmium was the cause. Itsuzo Shigematsu, who had once headed the group of researchers organized by the Japan Public Health Association (JPHA) that provided important results for the Ministry of Health and Welfare's announcement (see Sect. 7.4), was again made head of the group. He had introduced the concept of epidemiology research from the United States after World War II. Shigematsu, himself, was neutral regarding the cause of Itai-itai disease, but the research group continued to publish reports with ambiguous and inconclusive results, so that it lost its credibility in the end, even among other researchers (Matsunami 2010, pp. 318–319, 324–336).[30]

Itsuzo Shigematsu was born in Osaka in 1917. He graduated from the Medical School of Tokyo Imperial University in 1941 and studied public hygiene at the Harvard School of Public Health with Michio Hashimoto (see Sect. 7.4), receiving a Master's degree in Public Health in 1955. An interesting comparison can be made between the paths of the two Japanese who finished degrees at Harvard in 1955. Hashimoto, as the administrative official, played an important role in the early stages of anti-pollution and environmental measures by the Japanese Government.

---

[29] Takaya Kodama was born in Ashiya, Hyogo Prefecture, not far from Osaka. He graduated from Waseda University. In 1974, he was noticed because of an investigative report on Kakuei Tanaka, the Prime Minister of Japan. He died of cancer on May 22, 1975 (Sakagami 2003). Sakagami, who wrote Kodama's biography, sees Kodama's article on Itai-itai disease as the only negative thing in his writing career.

[30] See Toyama Broadcasting's documentary TV program "30 nen-me no gurei-zon: Kankyo-osen to kono kuni no katachi (in Japanese, 30th year's gray zone: Environmental pollution and the state of this country)," which criticizes the activity of Shigematsu's research group on Itai-itai disease and which received the Japan Congress of Journalists' Prize for 1999. This TV program was broadcast in January 1999.

Shigematsu, on the other hand, always directly or indirectly supported the government and thus mostly delayed anti-pollution measures, apart from when he worked for the Ministry of Health and Welfare. He chaired many governmental research committees on pollution-related or drug-induced diseases as an authority on epidemiology and seems to have played a key role in the presentation of inconclusive results and providing negative conclusions on the causes of the diseases. These negative roles did not seem to damage his career. On the contrary, he received every possible merit and honor.[31] He achieved high standing in exchange for allowing his "expertise" to be used for the maintenance of the status quo of the ruling system.

The difference between Hashimoto and Shigematsu can be explained by a certain aspect of academia. In the academic world, a researcher who advances a hypothesis that is in fact wrong may be criticized, whereas if the researcher advances no hypothesis at all, until the hypothesis is sure to be found correct in the end, no criticism occurs. In other words, a researcher is better off not to advance a hypothesis until conclusive results are available to support it. This "wait-and-see" attitude is praised, or at least accepted, in the academic world as the mark of a cautious researcher. However, this type of approach often creates trouble when tackling problems in technology-related social problems because of the untimely delays this approach causes when solving these kinds of problems.[32]

Nozomu Matsubara, a researcher in social statistics, once provided a statistical analysis of the damage this type of "wait-and-see" attitude can incur in the form of Type I and Type II errors (Matsubara 2002, pp. 196–200). Type I error, also known as a false positive, considers falsely negative things to be positive, so one does what one should not do. Type II error, also known as a false negative, considers falsely positive things to be negative, so one does not do what one should. With pollution problems, administrators commonly make Type II errors and experts promote this kind of error. Shigematsu, as an academic researcher, always tried his "best" not to make Type I errors, while ignoring the dangers of making Type II errors. In contrast, Hashimoto realized the danger created by the fear of making Type II errors and as an administrator, he took a different approach to pollution-related disease than did most scientists (Hashimoto 1988, p. 136).

---

[31] Shigematsu became a professor in the medical school at Kanazawa University, the Head of the Department of Epidemiology in the National Institute of Public Health in Japan, and the Chairman of the Radiation Effects Research Foundation, a research foundation that studies the effects of radiation in the survivors of the atomic bombings of Hiroshima and Nagasaki. He received the Sievert Prize for his work on Radio-epidemiology.

[32] Shigematsu chaired research committees on subacute myelo-optico-neuropathy (a drug-induced disease that occurred in the 1960s in Japan), Kawasaki Disease, Itai-itai disease, and Minamata disease. For the criticism of Shigematsu regarding Minamata disease, see Tsuda (2004, pp. 174–181).

7.7 A Pollution Prevention Program After the Trial: Collaboration of Experts and Citizens

On November 16, 1972, a group that included representatives of the victims' group, lawyers, and scientists inspected the Kamioka mine for the first time under the Pollution Control Agreement. The company's employees were more cooperative than they had been previously,[33] but some tension remained between the victims and the company. The second annual inspection, which took place from August 6th–8th 1973, was carried out more systematically. The inspection group made a list of survey items for each facility in the mine and demanded that the company prepare the necessary data beforehand. The group was divided into five subgroups, each in charge of inspection of a certain part in the mine. This second inspection became the prototype for later annual inspections (Matsunami 2010, pp. 344–346).[34]

Initially, and for several years, constant antagonism existed between the Residents' Association and the company. Each year, the Association carried out a full inspection of the mine and always in a strained atmosphere. The company broke its promise to improve its facilities more than once.

Also notable was the fact that local residents from polluted areas and environment-conscious people from all over Japan with various backgrounds (e.g., researchers, university professors, schoolteachers, and students) participated in the inspections.[35] The requirements of the Pollution Control Agreement have resulted in the performance of a full inspection of the mine every year since 1972. The present author participated in the 28th annual inspection of the mine in August 1999. By 2011, when the 40th annual inspection was held, the number of residents who had participated totaled 6,000, and the number of experts added up to 2,000 since the inception of the mine inspections (Hata 1998, p. 2; Hata 2011).[36]

---

[33] *Yomiuri Shinbun*, November 1972, Toyama edition pp. 12–13; *Yoyama Shinbun*, November 1972, p. 15.

[34] The annual inspection group is now divided into seven subgroups. On August 2010, the 39th annual inspection was held, with around 110 participants in the seven subgroups—including, for the first time, four officials from Toyama Prefecture. Kitanihon Broadcasting (KNB), a local broadcast network in Toyama Prefecture, reported the new on the visit on August 10, 2010.

[35] According to the Pollution Protection Agreement, persons whom the Residents' Association considers to be necessary can participate in the inspection; outside participants, including researchers and ordinary citizens, belong to this category.

[36] According to Akio Hata, who has been studying environmental problems in the Kamioka mine for a long time and who participated in various inspections of the mine in an earlier period, by 1998, the total number of residents who had participated totaled over 5,000, and the number of experts added up to 1,700 (Hata 1998, p. 2). By 2010, the number of participating residents reached 6,000 (Hata 2011). In 2011, Hata mentioned that the number of "experts" by 2011 was 1,000 (Hata 2011), but this number did not include lawyers. Adding the number of lawyers who participated in the inspections would bring the number of experts up to about 2,000 by 2011 (A. Hata pers. comm.).

Mitsui paid 280 million yen for residential inspections and additional research by outside experts from 1972 to 2010. The company has also invested 21.3 billion yen for pollution prevention measures based on the advice by experts of the Residents' Association in the same years (Hata 2011).

From 1974 to 1978, at the request of the Itai-itai disease Residents' Association, a research project on the mine and the river was conducted in cooperation of many universities in Japan. The project produced comprehensive reports on ways to reduce pollution[37] and proposed some measures for Mitsui's self-management to prevent pollution. The company itself started to issue an annual report on the preventive measures taken each year by the Kamioka mine against mining pollution.

The annual residential inspection of the mine was followed by a meeting between the mine representatives of the company and participants in the inspection, where the residents would ask questions and demand the adoption of various measures to prevent further pollution. In the early days, this meeting had a tense atmosphere and always lasted until almost midnight. However, after a few years, the atmosphere of the post-inspection meeting became friendlier and more sympathetic and a constructive relationship was formed with the company.

The attitude of the company gradually changed, for a number of reasons. For one thing, the experts of the Residents' Association, who were lawyers, university professors, and researchers in public institutions, cooperated with residents in the polluted areas and gained more expertise in pollution control than was possessed by the mining engineers employed by the company. Therefore, there Association's experts could give appropriate advice regarding improving the facilities. For example, during an additional inspection[38] conducted in 1977 these experts, together with members of the Residents' Association, a cadmium pollution source was found leading from the zinc smelting plant to the underground drains. The continuous efforts of outside experts helped the residents gain the trust of the company.

The second factor affecting the change in the company's attitude toward inspections was the change in Japanese society itself. During the 1980s, and especially in the 1990s, people became more environmentally conscious. Companies that were not

---

[37] The following five groups were organized in the project:

1. On the effluent from the Kamioka mine and refinery (Kyoto University)
2. On the smoke emissions from the Kamioka refinery (Nagoya University)
3. On the cadmium balance of the Kamioka refinery (University of Tokyo)
4. On the sedimentation and outflow of heavy metals into the Junzu River (Toyama University)
5. On the structural stability of the tailing dams at the Kamioka mine (Kanazawa University).

For the outline of the project, see Yoshida et al. (1999), p. 219.
Sometimes, the inspection by these experts would identify an unknown pollution source and appropriate countermeasures would be proposed (Kurachi et al. 1979, pp. 246–251).

[38] Besides the annual inspection, approximately ten additional inspections are held each year, based on specific themes like drainage, ventilation, pits, abandoned mines, planting (Hata 2011).

"green" could not survive: environmental consciousness appealed to the public[39] and was even profitable for companies. The Kamioka mine was one of the few mines in Japan that was sufficiently large and wealthy to continue its operations; it finally closed in June 2001. Today, the company uses the facilities remaining at Kamioka to refine and smelt ores imported from overseas mines.

The 40 years of continuous inspections has resulted in greatly reduced pollution outflow and improved mining facilities as well as in the development of an unprecedented cooperative relationship between the victims and the company. The total amount of discharged cadmium decreased from 35 kg/month in 1972 to 5 kg/month in 1997 and to 3.8 kg/month in 2010, and the mean concentration of cadmium in the effluent from the mine fell from 9 ppb in 1972 to 1.5 ppb in 1996 and 1.2 ppb in 2010. Improved dust collection reduced the total amount of cadmium discharged in smoke from more than 5 kg/month in 1972 to 0.4 kg/month in 1997 and 0.17 kg/month in 2010 (Hata 1998, p. 3; Hata 2011). The mean concentration of cadmium in the stream near the Kamioka mine should be at a level of 0.1 ppb (i.e., the background level) so that the outflow of cadmium from the a mine is negligible, thereby preventing any further pollution of the once polluted, but now restored, agricultural land. This aim was finally attained in 1996, when an almost negligible cadmium outflow was determined (Hata 1998, p. 7). The concentration of cadmium in the Ushi-ga-kubi canal, a major irrigation channel from Jinzu River was 0.7 ppb in 2011, the same as the background level (Hata 2011, p. 15).

7.8 Comparison Between the Cases of Itai-Itai Disease and Minamata Disease

Comparison of the case of Itai-itai disease with that of Minamata disease, another major pollution related disease from the 1960s, reveals four points that explain why Itai-itai disease was a rare example of successful pollution control in Japan: (1) socio-economical background of the patients in the local communities, (2) related laws, (3) the pollution protection agreement, and (4) participation by experts and citizens.

Most of the patients who suffered from Itai-itai disease were peasant women from the rich rice field area. Kin-ichi Yoshioka (see Sect. 7.3), one of earliest researchers on the cause of the disease, analyzed the socio-economic structure of patients with Itai-itai disease and noted that Kumano village (now the Kumano District of Fuchu-machi in Toyama Prefecture), the most-polluted area with a high concentration of patients, was also the area with the largest per-farm acreage in Toyama Prefecture. More patients came from farms of one hectare or more than came from farms with less than one hectare of land (Yoshioka 1970, p. 70).

---

[39] "Environment" is always one of the most important key words in the message from the heads of the company on the internet site of the Mitsui Mining and Smelting Co. Ltd. (Message from Management, accessed August 2, 2014, http://www.mitsui-kinzoku.co.jp/en/company/c_message/).

Kumano village was located only a one-hour bus ride from Toyama city, the central city of Toyama Prefecture. Most farmers were full-time and their management was stable, and malnutrition was unthinkable even that soon after World War II (Yoshioka 1970, p. 73).

In contrast, Minamata disease struck the poor fishermen who lived on the periphery of the local socioeconomic structure. Masazumi Harada (1934–2012), famous for his research on Minamata disease, once called this situation "double discrimination in public and private." As he explained, "in addition to the discrimination by the authority, patients were discriminated by ordinary people, workers and citizens, even by other fishermen." "People often referred to patients as "those guys," meaning something different from "ours" with connotation of poor by nature, strangers, lower class. For ordinary citizen in Minamata city, Minamata disease was the disease of the fishermen, who ate queer fish caught in the sea" (Harada 1989, pp. 22–23). A movement arose in Minamata city in 1972 to change the name "Minamata disease," because of damage to the image of Minamata city; 72 % of eligible voters in the city approved the change (Minamata City 2007, p. 9).

One must also note, in the case of Itai-itai disease, the application of the mining law, which permits liability without fault. Therefore, the plaintiffs did not have to prove the company's fault and intent (Matsunami 2010, pp. 112–113). All they had to do was to prove that the Kamioka mine discarded waste polluted with cadmium into the Jinzu River. In contrast, in the case of Minamata disease, civil law was valid, which meant that the plaintiffs had to prove that the defendant company purposely caused the damage. This made the Minamata case very difficult to win. The first lawsuit on Minamata disease, which started at almost the same time as that of Itai-itai disease, took 4 years and 9 months to the first verdict (from June 1968 to March 1973), while that of Itai-itai disease took 3 years and 3 months (from March 1968 to June 1971).

The Pollution Control Agreement, concluded between Mitsui and the victims' group after the Itai-itai disease lawsuit, had epoch-making content. As shown in the previous section, the 40-year continuous inspections under the agreement has now restored the concentration of cadmium in the Jinzu River to the natural background level. However, this type of agreement was not reached in the case of Minamata disease.

The importance of the involvement of experts and citizens after the lawsuits must also be recognized. In the case of Itai-itai disease, victim residents and concerned citizens from all over the country, as well as experts like lawyers and scientists, have continuously participated in the annual inspections under the Pollution Control Agreement. In the case of Minamata disease, even though doctors in the medical school at Kumamoto University, a local national university, played an important role in the early stage in identifying the cause of the disease and in helping patients at the early stage, most of these doctors were not actively involved with the problems later on. Patients and victims' organizations were isolated in Minamata city. The victims' organizations were also sometimes divided by their political positions.

7.9 Conclusion and Some Suggestions for Environmental Regeneration in Fukushima

The present study describes the long and tortuous path by which Itai-itai disease became recognized. Second, it discusses how Itai-itai victims and concerned residents, in collaboration with their supporters such as lawyers and experts, fought a difficult battle to obtain compensation. Third, and most importantly, this struggle resulted in the establishment of a comprehensive agreement between the victims, concerned residents, and the polluting company, which allowed concerned residents, with the help of supporting experts, to participate in decision making and pollution prevention, thereby offering a long-term solution to industrial pollution. The experience itself is valuable because it exemplifies a rare case of successful pollution control in Japan.

The Itai-itai disease case also offers good materials for analysis of the role of experts in addressing technology-related social problems such as pollution control.

This present case study can contribute to the understanding of the nature of expertise. The 1995 study by Stephen Epstein analyzed "lay experts" on AIDS, who themselves were potential patients, or friends or family of patients, and who devoted much energy to understanding the disease (Epstein 1995). In the case of Itai-itai disease, some lawyers, such as Jun-ichi Matsunami, became "lay experts" of a sort, who (although not physicians or scientists) gained more knowledge about the disease and pollution than did experts in company's employ. These lawyers became "nonscientist-but-nonetheless-knowledgeable participants" in the process (Evans and Collins 2008, p. 611).[40]

The current study also points out the significance of combining different types of expertise. The cooperation between Noboru Hagino, a local physician, Kin-ichi Yoshioka, an agronomist, and Jun Kobayashi, an analytical chemist, helped to uncover the cause of the disease.

However, other experts, like Jugoro Takeuchi and Itsuzo Shigematsu, directly or indirectly supported the position of offending companies and the apathetic government, thereby delaying anti-pollution measures. As Evans and Collins pointed out, these experts spoke as experts and also as political agents (Evans and Collins 2008, p. 612). In case of Shigematsu, his attitude can be explained by a certain aspect of academic world, as shown in the Sect. 7.6. His "wait-and-see" attitude would be praised, or at least accepted, in the academic world as the mark of a cautious researcher, whereas in the real world it caused much damage. Like Nozomu Matsubara (Matsubara 2002, pp. 196–200), this damaging "wait-and-see" attitude can be analyzed in the terms of statistics, i.e., as Type I and Type II errors.

The case of Hashimoto, then the head of the pollution department of the Ministry of Health and Welfare, is also important. Hashimoto insisted, as an administrator, that public administration should act based on the available scientific evidence,

---

[40] Regarding "lay experts," see also, for example, Brown (1992), Callon (1999), Callon et al. (2001).

rather than waiting for conclusive scientific proof, in order to avoid an untimely decision (Hashimoto 1988, p. 136). This is obviously the first major instance of the precautionary principle in Japan.

The case of Itai-itai disease was a good example of public participation in societal problems. The public participation in the annual inspections of the mine played a significant role in the reduction of cadmium concentrations in the Jinzu River to a safe level.[41] As this chapter shows, this public participation was realized largely due to the Pollution Protection Agreement, which was established after the lawsuit. This agreement was significant, but it is also the only example of its kind in Japan. Therefore, one can learn a lot from the unique case of Itai-itai disease. Japan has suffered again from serious environmental pollution caused by the Fukushima Daiichi nuclear disaster following the 2011 Great East Japan Earthquake on March 11, 2011.[42] Recovery from this disaster is predicted to take a long time, just like the pollution problems of the 1960s. If the pollution is not managed properly, the recovery may become bogged down, as in the case of Minamata disease. The lesson revealed by the case of Itai-itai disease, in the opinion of this author, may provide a way to avoid the situation that occurred in Minamata.

The author also thinks that the following three points suggested by the experience of Itai-itai disease in Toyama Prefecture for "environmental regeneration" could be directly applicable to the restoration and enhancement of areas degraded by the nuclear power plant disaster in Fukushima Prefecture.[43]

First, environmental regeneration takes time. On December 17th, 2013 the victims' association and Mitsui, the polluter company, reached their final agreement on the complete resolution of the cadmium pollution damage caused by the Kamioka mine (Hata and Mukai 2014, pp. 12–33). It took more than 100 years after the appearance of the first patients and about 40 years after the victory of the lawsuit for the polluter company to complete the environmental regeneration in terms of river water, agricultural lands, and human health.[44] This implies that a very long time will also be needed to restore the damaged lands in Fukushima.

Secondly, the Itai-itai disease experience has suggested that collaboration between experts and concerned residents would be needed to obtain the desired results. Yuko Fujigaki, a researcher in science and technology studies (STS), stated that scientists often feel responsibility only for scientific communities and have paid

---

[41] Itai-itai disease is not a thing of the past, however. New patients with Itai-itai disease continue to be found even now. Patients exposed to cadmium in their youth show continued disease development as they get older. See Fujikawa (2005, pp. 112–113) and Aoshima (2004).

[42] On the Earthquake, see the relevant site of Japan Meteorological Agency (http://www.jma.go.jp/jma/en/2011_Earthquake/2011_Earthquake.html, assessed August 7, 2014).

[43] Akio Hata, an environmental scientist, who has been studying the Itai-itai disease problems last 40 years, has proposed several suggestions for the solution of the Fukushima nuclear disaster (Hata and Mukai 2014, pp. 226–230).

[44] Even though there are still people suffered from damage from cadmium, the final agreement has promised the establishment of the health supporting system for the residents at Jinzu River area (Hata and Mukai 2014, pp. 16–21).

less attention to society outside of their community (Fujigaki 2003, pp. 25–28). The solution of environmental problems like the cases of Itai-itai disease and the Fukushima nuclear disaster will require that experts stay aware that what the public expects is different from what the scientific community expects. This is a very important concept, since the nuclear expert community is often described as "Japan's Nuclear Power Village."

Thirdly, as a consequence of these first two points, the education and training of future specialists in the field of nuclear industry becomes problematic because the need for highly qualified experts will continue for the next 100 years for both stopping and regenerating the industry. Chemistry and the chemical industry suffered some setbacks in the 1970s and 1980s because of the pollution problems that arose in the 1960s and 1970s. Now nuclear science and technology are having their turn.

One should study the experience of "Toyama" to plan the regeneration of "Fukushima."

References

Aoshima, K. (2004). Jinzu-gawa ryuiki jumin no kadomiumu bakuro to jin-shogai: Genjo to korekara (in Japanese, Cadmium Exposure and Kidney Disorder of Jinzu River Residents: Present and Future). Itai-ita byo seminar koen-shu (Proceedings of Itai-itai disease Seminar), Vol. 22, pp. 6–23.
Brown, P. (1992). Popular epidemiology and toxic waste contamination: lay and professional ways of knowing. *Journal of Health and Social Behaviour, 33,* 267–281.
Callon, M. (1999). The role of lay people in the production and dissemination of knowledge. *Science, Technology and Society, 4*(1), 81–94.
Callon, M., Lascoumes, P., & Barthe, Y. (2001). *Agir dans un monde incertain.* Paris: Éditions du Seuil.
Daintith, J. (Ed.). (2008). *Oxford dictionary of chemistry* (6th ed.). Oxford: Oxford University Press.
Dehn, U. (1995). *Tanaka Shōzō: ein Vorkämpfer für Menschenrechte und Umweltschutz.* Tokyo: Deutsche Gesellschaft für Natur- und Völkerjunde Ostasisens (OAG).
Epstein, S. (1995). The construction of lay expertise: AIDS activism and the forging of credibility in the reform of clinical trials. *Science, Technology and Human Values 20,* 408–437.
Evans, R., & Collins, H. (2003). Expertise: from attribute to attribution and back again? In E. J. Hackett, O. Amsterdamska, M. Lynch, & J. Wajcman (Eds.), *The handbook of science and technology studies* (3rd ed., pp. 609–630). Cambridge: The MIT Press.
Fujigaki, Y. (2003). *Senmon-chi to kokyo-sei: Kagaku-gijutsu-shakai ron no kouchiku ni mukete* (in Japanese, *The Public Ethic and the Spirit of Specialism*). Tokyo: University of Tokyo Press.
Fujikawa, K. (2005). Kogai Higai Houchi no Sho-yoin (in Japanese, Various Factors for Negligence to Pollution). *Kankyo-shakai-gaku Kenkyu (Journal of Environmental Sociology), 11,* 103–116.
Hagino, N. (1968). *Itai itai byo tono tatakai* (in Japanese, *Fight Against Itai-itai Disease*). Tokyo: Asahi Shinbun Sha.
Harada, M. (1989). *Minamata byo ga utsusu sekai* (in Japanese, *The World, Reflected by Minamata*). Tokyo: Nihon Hyoron Sha.
Hashimoto, M. (1988). *Shi-shi Kankyo Gyosei* (in Japanese, *Personal Reminiscence on Japanese Environmental Administration*). Tokyo: Asahi Shinbun Sha.

Hata, A. (1998). Itai-itai byo saiban go no jumin-sanka ni yoru hassei-gen-kisei to kigyo-joho-kokai no yakuwari (in Japanese, the control of a pollution source by the participation of local residents and the role of access to the involved company's information after the Itai-itai disease lawsuit). *Mizu-shigen Kankyo Kenkyu (Journal of Water and Environmental Issues), 11*, 1–10.

Hata, A. (2011). Kamioka kozan no haisui taisaku no totatsuten to kongo no kadai (in Japanese, Accomplishment and further problems in the drainage measures in Kamioka mine). Proceedings of the Symposium to commemorate 40th annual inspection of the Kamioka mine, Toyama, Japan (pp. 13–17), August 6th, 2011.

Hata, A., & Mukai, Y. (2014). *Itai-itai byo to Fukushima: koremade no 100 nen korekara no 100 nen* (in Japanese, *Itai-itai disease and Fukushima: previous 100 years and next 100 years*). Tokyo: Goto Shoin.

Hatta, S. (1983). *Shi no kawa to tatakau: itai-itai byo wo otte* (in Japanese, Fighting against Death River: following Itai-itai disease) (ser. Kaisei-sha bunko). Tokyo: Kaisei-sha.

Isono, Y. (1999). Itai-itai disease and the Pollution Control Agreement. In K. Nogawa, M. Kurachi & M. Kasuya (Eds.), *Advances in the prevention of environmental cadmium pollution and countermeasures: Proceedings of the International Conference on Itai-itai disease, Environmental Cadmium Pollution and Countermeasures, Toyama, Japan*. 13–16 May, 1998 (pp. 213–214). Kanazawa: Eiko Laboratory.

Itai-itai byo sosho bengo-dan (the defense counsel of Itai-itai disease suit), Ed. (1972). *Itai-itai byo saibain* (in Japanese, *Itai-itai disease lawsuit*) Vol. 3. Tokyo: Sogo Tosho.

Iwamoto, A. (1999). Restoration of Cd-Polluted Paddy Fields in the Jinzu River Basin. In K. Nogawa, M. Kurachi & M. Kasuya (Eds.), *Advances in the prevention of environmental cadmium pollution and countermeasures: Proceedings of the International Conference on Itai-itai disease, Environmental Cadmium Pollution and Countermeasures, Toyama, Japan*, 13–16 May, 1998 (pp. 179–183). Kanazawa: Eiko Laboratory.

Kaji, M. (2005). Itai-itai byo mondai kaiketsu nimiru senmonka to shimin no yakuwari (in Japanese, The role of expert and citizen participation in Pollution Control: in the case of Itai-itai disease). In Y. Fujigaki (Ed.), *Kagaku-gijutsu-shakai-ron no giho* in Japanese, (*Case Analysis and Theoretical Concepts for Science and Technology Studies*) (pp. 21–42). Tokyo: University of Tokyo Press.

Kaji, M. (2009). Shakai ni okeru senmonka no yakuwari: itai-itai byo no byoin wo meguru giron kara (in Japanese, The role of experts in society: the debates on the cause of Itai-itai disease). In M. Kaji, M. Saijo, & K. Nohara (Eds.), *Kagaku-gijutsu komyunikeishon nyumon* (in Japanese, *An Introduction to Science and Technology Communication*) (pp. 73–86). Tokyo: Baifu-kan.

Kaji, M. (2012). Role of experts and public participation in pollution control: the case of Itai-itai disease in Japan. *Ethics in Science and Environmental Politics, 12*, 99–111.

Kankyo-cho (Environmental Agency in Japan). (1973). *Showa 48 nen ban Kankyo Haku-sho* (in Japanese, *The White Paper on the Environment in Japan for 1973*). Tokyo: Ookura sho Insatsu-Kyoku (Printing Bureau of Ministry of Finance).

Kodama, T. (1975). Itai itai byo wa maboroshi no kogai-byo ka? (in Japanese, Is Itai-itai disease a mere phantom?) *Bungei-shinju, 53*, 312–338.

Kurachi, M. (1999). General research into cadmium poisoning prevention in the Jinzu river basin and the worldwide significance of pollution-free mining. In K. Nogawa, M. Kurachi & M. Kasuya (Eds.), *Advances in the prevention of environmental cadmium pollution and countermeasures: Proceedings of the International Conference on Itai-itai disease, Environmental Cadmium Pollution and Countermeasures, Toyama, Japan*, 13–16 May, 1998 (pp. 149–154). Kanazawa: Eiko Laboratory.

Kurachi, M., Tonegawa, H., & Hata, A. (1979). *Mitsui Shihon to Itai-Itai Byou* (in Japanese, *Mitsui Capital and Itai-itai disease*). Tokyo: Ootsuki-Shoten.

Matsubara, N. (2002). Kankyo-gaku ni okeru data no jubunsei to ishi-kettei-handan (in Japanese, Adequacy of data and decision-making judgment in environmental studies). In H. Ishi (Ed.), Kankyo-gaku no gihou (in Japanese, *Social Methods of Environmental Studies*) (pp. 167–214). Tokyo: University of Tokyo Press.

Minamata City. (2007). *Minamata byo: sono rekishi to kyokun* (in Japanese, *Minamata disease: Its history and lessons*). Minamata city.

Matsunami, J. (1998). *Aru Hantai Jinmon* (in Japanese, *Cross-examinations*). Tokyo: Nihon Hyoron.

Matsunami, J. (2010). *Kadomiumu higai hyaku nen: Kaiko to tenbo* (in Japanese, *A Hundred years of Cadmium Poisoning: Recollection and Prospects*). Toyama: Katsura Shobo.

Nicaud, P., Lafitte, A., & Gros, A. (1942). Les troubles de l'intoxication chronique par le cadmium. *Archives des maladies professionnelles hygiene et toxicologie industrielles, 4*, 192–202.

Nixon, R. (1971). *Public papers of the Presidents of the United States: Richard Nixon, containing the public messages, speeches, and statements of the President 1970*. Washington D.C.: United States Government Printing Office.

Sakagami, R. (2003). *Munen wa chikara Densetsu no ruporaita Kodama Takaya no 38 nen* (in Japanese, *Shame is Power: Takaya Kodama, Legendary writer's 38 years*). Tokyo: Joho-Center-Shoppankyoku.

Sismondo, S. (2010). *An introduction to science and technology studies* (2nd ed.). Chechester: Wiley-Blackwell.

Taketani, M. (1957). *Gen-sui-baku Jikken* (in Japanese, *Nuclear Testing*). Tokyo: Iwanami shoten.

Tsuda, T. (2004). *Igaku-sha wa Kogai-jiken de nani wo shitekita-noka?* (In Japanese, *What did medical researchers do in pollution cases?*). Tokyo: Iwanami Shoten.

Yamada, Y. (1951). On the lead mining and metallurgy of Japan (in Japanese). 1951. *Nihon Kogyo Kaishi (Journal of the Mining Institute of Japan), 67*, 46–51.

Yoshioka, K. (1961). Jinzu-gawa suikei kogai kenkyu hokokusho (in Japanese, Report on mining pollution study of Jinzu River and its tributaries), self-publishing. Later published in Yoshioka, 1971, 5–95.

Yoshioka, K. (1964). Itai-itai byo to kogai tono kanrensei ni tsuite no eikigaku teki kenkyu (in Japanese, Epidemiological study on relationship between Itai-itai disease and mining pollution). *Yomaguchi Igaku (Yamaguchi Medical Conference Journal), 3*, 146–170.

Yoshioka, K. (1970). *Itai-itai byo kenkyu: kadomiumu nogyo-kogai kara ningen kogai (itai-itai byo) he no tsuikyu* (in Japanese, *Studies on Itai-itai disease: from cadmium agricultural pollution to pollution to human being (Itai-itai disease)*). Yonago, Tottori Prefecture: Tatara Shobo.

Yoshida, F., Hata, A., & Tonegawa, H. (1999). Itai-itai disease and the countermeasures against cadmium pollution by the Kamioka mine. *Environmental Economics and Policy Studies, 2*, 215–229.

Yui, M. (1984). *Tanaka Shozo* (in Japanese). Tokyo: Iwanami Shoten.

Chapter 8
The Monju Trial: Nuclear Controversy in Japan

Tadashi Kobayashi and Minako Kusafuka

Abstract This chapter analyses an administrative lawsuit from 2003 concerning the Monju prototype fast-breeder reactor in Fukui as an example of nuclear disputes in Japan, and explores what it means for the safety of a nuclear reactor to be disputed in a judicial court. The analysis reveals that such litigation concerning a nuclear facility focuses on the validity of administrative procedures and does not pass judgement on the safety of the nuclear reactor itself. With regard to judgement on the safety of nuclear reactors, the views of the administrative authorities' experts, who exercise *engineering judgement*, which inevitably involves a degree of uncertainty, are treated with great respect. In administrative litigation, the stepwise regulation of installation and operation of nuclear power plants is taken as a premise, and matters such as the final disposal of nuclear waste and the methods of reprocessing spent nuclear fuels are beyond the scope of judicial review, as are the social aspects of utilising nuclear energy. The analysis highlights the limitations of administrative lawsuits to solve social conflicts about nuclear power plants in Japanese society.

8.1 Introduction

This chapter analyses an administrative lawsuit concerning the Monju prototype fast-breeder reactor in Fukui as an example of nuclear disputes in Japan. There are several reasons why this study focuses on the Monju trial.

First, this lawsuit is a typical trial in that science and technology were disputed in a judicial court. While science and technology (or technoscience) greatly expanded in the 20th century and benefitted humans tremendously, they also produced negative by-products, such as diverse forms of pollution, medical scandals, and global

---

T. Kobayashi (✉)
Osaka University, Osaka, Japan
e-mail: kobayashi@cscd.osaka-u.ac.jp

M. Kusafuka
The University of Tokyo, Tokyo, Japan
e-mail: ckonami@mail.ecc.u-tokyo.ac.jp

environmental risks. Consequently, science and technology have become sources of various sorts of social dispute. For any society, social conflicts are inevitable, and thus every society has developed systems to handle them. No one would doubt that, in modern societies, the judicial system is one of the principal mechanisms for resolving disputes, and science and technology issues are brought under the scrutiny of this system in order to be resolved.

When science and technology are on trial, peculiar difficulties arise in when the views of *experts* are disputed. In such cases, the kinds of resolutions that judicial decisions can provide or should provide are an issue that demands thorough consideration. This kind of problem is likely to increase.[1] How the judicial system, as a conventional mechanism for resolving social conflicts, will manage to deal with the continuing expansion of science and technology after the 20th century poses a serious challenge to us living in the 21st century (Jasanoff 1995).

Second, the case of the Monju trial presents a good example of problem setting that demands a broader perspective, namely, the interface between science/technology and the public. Because the construction of other nuclear power plants or a fast-breeder reactor like Monju are planned as part of the national long-term energy policy, and responsibility for the installation plans rest with the government. In this sense, nuclear plant construction projects are public projects. The Monju trial features conflict between the interests of local residents and the public interests of a nation and thus the decision in the Monju trial affects not only the installation of a fast-breed reactor at Monju but also the nation's long-term energy policy as a whole.[2] Although conflicts of a local-versus-national nature arise most prominently in the siting of nuclear plants, they can be also found in many other issues.

The issues demonstrated here are the basic dilemmas of politics: how to resolve the conflicting interests of a region and a nation with regard to the siting of a public facility, and what kind of decision-making process should be employed. Rephrasing these problems more explicitly in relation to science and technology, we are concerned with how society should understand sciences and technologies that are expected to benefit the public welfare, and how to make a societal decision to utilise them. In short, it is a matter of how we govern science and technology in society.

With these problems in mind, the following sections examine the Monju trial as a case study to explore what it means for the safety of nuclear facilities to be disputed in court. First, the background of this lawsuit is outlined, and then the

---

[1] The original version of this paper in Japanese was written before 11 March 2011. Currently (2014), there are over 20 cases before the courts in Japan concerning nuclear power plants. In 2014, Fukui District Court decided that Kansai Electric Company should not resume operation of the Oi nuclear power plant. We cannot judge at this point whether this is the beginning of a process of change or an exceptional case in the Japanese judicial system.

[2] As Monju has been shut down at the time of writing, Japan's national plan for nuclear fuel reprocessing is tentatively forced to rely on the *pluthermal* (plutonium for thermal use) project, which aims to burn a mixed oxide form of plutonium and uranium in plutonium-thermal reactors. The *pluthermal* project has also been criticised for many reasons. Hitoshi Yoshioka, who served as an expert adviser of the Atomic Energy Commission of Japan and has actively studied and spoken about atomic energy use in Japan, gave a thorough account of the *pluthermal* project (Yoshioka 1999).

media's coverage of the lawsuit is described because it plays a major role in framing the issue for society. Second, the article reviews various players regarded as important for this analysis, and then explicates how each player reacted to the lawsuit. Third, by examining the judgements of the Monju trial and other precedents, we discuss the issues that need to be addressed from the standpoint of science and technology studies (STS). Fourth, the article looks into the views of jurists regarding the Monju trial and other trials in which nuclear facilities were the subject of dispute. Lastly, we lay out the limitations of the judicial system as a principal mechanism for conflict resolution of issues involving science and technology.

8.2 Outline of the Monju Trial

On 27 January 2003, the Kanazawa branch of the Nagoya High District Court ruled the nullification of the Monju construction permit issued in 1983.[3] This decision overturned the decision of the Fukui District Court that ruled in favour of the government. This judgement of 2003 was the first time the government or the licensee of nuclear facilities had lost a suit, and the decision had a powerful impact on groups both for and against the installation at Monju. Monju is a MOX[4]-fuelled, sodium-cooled, loop-type reactor with electric-generating power of 280 MW. A chronology of the events regarding the installation at Monju and the lawsuits is summarised in Table 8.1.

The Japanese government started to develop the installation plan for Monju in the 1970s. From the beginning of this planning, opposition movements against the installation plan arose. In 1985, 40 local residents who were members of a group

---

[3] In order to request the cancellation of an administrative disposal, at the time of the Monju trial, a plaintiff had to file a case within three months of a disposal being acknowledged (Article 14 of the Administrative Litigation Act, 'statute of limitations'). In the case of the Monju trial, as it was already past this period when the plaintiffs tried to file a lawsuit, they had to file a lawsuit to request nullification (invalidity confirmation) of the construction permit. While a suit for invalidity confirmation is not accompanied by any statute of limitations, a plaintiff must establish *material and clear* illegality (deficit) concerning an administrative disposal. This requirement was established by the Supreme Court third petty bench ruling on 26 December 1995, as a precedent. This ruling confirmed that, 'even if an administrative disposal is illegal, except a case in which the illegality is confirmed as so material and clear that the disposal must be nullified by necessity, the disposal shall remain in full force unless it is revoked lawfully'. This requirement became the point at issue in the High Court decision on the Monju trial. The High Court judged, as the potential dangers of the administrative disposal came under *the extraordinary situation* in this case, that it was enough to establish *material* illegality only, and not necessary to establish *clear* illegality. Meanwhile, a revision to the Administrative Litigation Act of 2004 extended the statute of limitations for filing an administrative lawsuit to request a cancellation of an administrative disposal to six months.

[4] MOX fuel is mixed oxide nuclear fuel that is made by blending uranium with reprocessed plutonium taken from spent nuclear fuels. The International MOX Assessment (IMA) project cautioned in its final report that the use of MOX fuel in light water reactors would raise special safety, security, and economic considerations (Takagi et al. 1997).

Table 8.1 Chronology of events related to the Monju trial

Date	Events
1983, 5	Government issued the construction permit for Monju
1985, 9	Local residents filed two lawsuits (administrative and civil) in Fukui District Court to nullify construction permit and to stop construction
1985, 10	The Monju construction work started
1987, 12	Fukui District Court rejected the plaintiffs' claim, saying that local citizens were not eligible for plaintiff status. The plaintiffs appealed the decision to the High Court
1989, 7	Kanazawa branch of the Nagoya High District Court recognised the qualification of plaintiffs who lived within 20 km of Monju. Both the plaintiffs and the defendant appealed the decision to the Supreme Court
1992, 9	The Supreme Court fully recognised all of the plaintiffs' qualifications and ordered that the case be sent back to the lower court
1992, 10	The Supreme Court rejected an appeal filed by the citizens seeking the cancellation of permissions for the Ikata Unit 1 and the Fukushima II Unit 1 nuclear plants (administrative lawsuit)
1994, 4	Monju went critical for the first time
1995, 5	Monju began to provide electricity
1995, 12	Sodium leaked out from the Monju
2000, 3	Fukui District Court rejected the plaintiffs' claims for both the civil and administrative cases. The plaintiffs appealed the decision to the High Court
2001, 6	The Japan Nuclear Cycle Development Institute applied for permission for the design modification of Monju to restart operations
2002, 12	The Ministry of Economy, Trade, and Industry (METI) permitted the modification
2003, 1	The Kanazawa branch of the Nagoya High District Court ruled the nullification of the Monju construction permit issued in 1983. The defendant appealed the decision to the Supreme Court
2004, 12	The Supreme Court accepted the defendant's appeal

Source CNIC (2003). Table modified by authors

called Fukui Prefecture Citizens against Nuclear Power filed two lawsuits in Fukui District Court. One was administrative litigation against the prime minister (the government) to nullify the construction permit for Monju. The other was a civil lawsuit against the licensee, the Power Reactor Nuclear Fuel Development Corporation (PNC), which was renamed the Japan Nuclear Cycle Development Institute (JNC) in 1998, to seek injunction of the construction and operation. The Fukui District Court separated the two lawsuits, rejecting the local residents' claim in the administrative lawsuit, saying that they were not eligible for plaintiff status (Fukui Prefecture Citizens against Nuclear Power 1996).[5] This decision was reversed in the

---

[5] At the first trial, the government (the defendant) claimed that a court should not judge the illegality of a construction permit of an administrative disposition and thus should dismiss this suit. The Fukui District Court accepted the government's claim. In the second instance, the High Court admitted this lawsuit but rejected the status of plaintiffs who lived more than 20 km away from Monju. The Supreme Court reversed the decision of the High Court, however, and recognised the qualifications of all the plaintiffs. Meanwhile, the nuclear meltdown at Chernobyl took place in 1986, a year after the local residents filed this case.

first and second appeal courts. In 1992, the Supreme Court fully admitted all of the plaintiffs' qualifications and ordered that the case be sent back to the Fukui District Court. Fukui District Court then started the inquiry over, in parallel with the ongoing civil lawsuit.

Also in 1992, the Supreme Court rejected appeals filed by citizens seeking the cancellation of permissions for the Ikata Unit 1 and the Fukushima II Unit 1 nuclear plants (administrative lawsuits). These decisions determined the basic pattern of nuclear-related lawsuits in Japan thereafter, which we will come back to later.

During the trials at the Fukui District Court, Monju went critical for the first time, in 1994, and in August 1995, it started to transmit electricity. In December 1995, however, approximately 0.7 tons of sodium leaked from the secondary cooling system, reacted with oxygen in the air, and caused a fire. The leaked sodium also damaged the floor's steel liner for prevention of contact between sodium and moisture in the concrete below. Soon after the accident, it was also revealed that PNC had tried to cover up the seriousness of the accident and the resulting damage, provoking massive public outrage. After the accident, the reactor was forced to stop operations.

The Fukui District Court closed the two lawsuits (administrative and civil) in April 1999 and pronounced judgements in favour of the defendants in March 2000. The plaintiffs appealed both decisions to the Nagoya High District Court, and its Kanazawa branch started the trials. The High Court examined the administrative litigation first. For this examination, the court held few trial hearings but 18 scheduling conferences,[6] which resulted in the judgement to nullify the construction permit on 23 January 2003. After this decision was pronounced, the plaintiffs dropped the civil suit, with the agreement of the defendant, but the defendants who lost the administrative suit appealed the decision to the Supreme Court on 31 January 2003. The Supreme Court accepted the appeal on 2 December 2004, and decided to hold oral proceedings in March 2005. The first petty bench of the Supreme Court quashed the High Court's decision and dismissed the plaintiffs' appeal.

The rest of this chapter focuses on the decision of the Kanazawa branch of Nagoya High District Court in 2003, rather than on the decision of the Supreme Court in 2005. This is because the High Court's decision of 2003 had anomalous aspects as a decision of administrative litigation, while the Supreme Court's decision could be anticipated to some extent in light of conventional legal theories. Analysing this unusual decision, therefore, sheds light on the important issues that need to be addressed regarding the characteristics and ways to resolve conflicts about nuclear risks in Japan.

---

[6] As the defendant, the government was unhappy with this process of examination. At the same time, the government showed repentance for having downplayed the scheduling conferences.

8.3 Media Coverage: The Newspapers' Framing[7]

As it was unprecedented that the local residents won a lawsuit concerning nuclear plants, the High Court's decision was widely reported in the news media in Japan. All of the national newspapers explained the decision and reported comments by the involved parties in the Monju trial, including the defendant (the government), the licensee (JNC), and the plaintiffs (local residents), as well as key intellectuals. On the day following the decision, 28 January, editorial articles in the morning editions of all the national newspapers discussed it. The newspapers focused on a number of points in their coverage of the trial.

The analysis presented here examines a table that contrasts the claims of each party involved in the trial. Both *The Mainichi Newspapers* and *Asahi Shimbun* used such a table in their articles. This analysis uses a table from *The Mainichi Newspapers* because it includes information that is more detailed.

The three points of issue listed in Table 8.2 constitute almost the same points that were disputed in the trial. We say 'almost' because one important point is missing: that specific to the administrative litigation, referred to as 'invalidity confirmation'.[8] In the Monju administrative litigation, the plaintiffs sought not the cancellation, but the nullification of the construction permit. Putting this in legal terms, the suit was for the invalidity confirmation of the administrative disposal. In such suits, the plaintiffs must establish the *material and clear illegality* (or deficit) concerning the administrative disposal. In the case of the Monju trial, then, the plaintiffs had to establish that the process of the government's verification and judgement of the construction permit contained material mistakes or deficits. This point was hardly mentioned in the newspaper articles because it could be regarded as just a technical matter inherent in administrative litigations. This, by itself, typifies the media coverage of the Monju trial. By contrast, as the defendant, the government focused on this point and used it as the major reason for the appellate as we shall discuss later.

On these three points of issue that the newspapers reported, at first glance it is clear that they concern whether the possibility of accidents could be anticipated. These accidents were so-called low-probability events. While the plaintiffs argued for the possibility of their occurrence, the defendant argued that their probability was extremely low and that adequate preventive measures had already been taken. The litigation can therefore be interpreted thus: on one hand, the lower court supported the defendant's claim, while on the other hand the High Court was in favour of the plaintiffs' claim. Most of the Japanese newspapers reported along these lines. In other words, their articles understood the High Court's findings

---

[7] All the descriptions of newspaper articles in this section are based on the Nagoya edition, dated 28 January 2003.

[8] See footnote 2.

Table 8.2 Major arguments of the Monju Trial

1. Preventive measures against sodium leakage	
Plaintiff *local residents*	The experiments after the accident revealed that a floor's steel liner, intended to prevent an explosion caused by contact between sodium and moisture in the concrete below the floor, could be perforated by corrosion. After these experiments, the JNC applied its remodelling plan of Monju to include preventive measures against sodium leakage. This change itself demonstrates critical flaws in the initial safety review of the baseline design
Defendant *government*	Although the floor's steel liner was corroded by the sodium leakage accident, it did prevent direct contact between sodium and concrete. The baseline design of utilising a steel liner to prevent their contact was valid. PNC's application for remodelling the baseline design was only to improve safety
Fukui district court's judgement	Although the possibility of a floor liner being corroded could not be denied, if the detailed design made after the safety review could prevent contact between sodium and concrete, the validity of the baseline design using a steel liner to prevent contact would not be lost
High court's judgement	There were critical flaws and errors in the safety review that overlooked the possibility of corrosion of the floor liner. The current system does not ensure the prevention of direct contact between sodium and concrete. Consequently, the possibility of all cooling systems losing their functions is undeniable
2. Preventive measures against damaging steam generator tubes	
Plaintiff *local residents*	As an accident in the UK revealed, if water or steam leaked out from steam generator tubes, it could cause an explosive reaction with the leaked sodium. This would result in the phenomenon of *high-temperature rupture* that has the potential to cause many tubes to rupture simultaneously in a chain reaction. It was a critical flaw of the safety review that it overlooked the possibility of this phenomenon
Defendant *government*	The preventive measures, such as the installation of a manometer in the steam generators to detect ruptures of tubes at an early stage to remove water and steam swiftly, were adequate. A high-temperature rupture would not occur
Fukui district court's judgement	As the experiments at PNC and calculations indicate, it is difficult to anticipate that a high-temperature rupture would occur at Monju. Even if the safety review overlooked this possibility, it cannot be said that the review lacked rationality
High court's judgement	For the steam generators of Monju, the possibility of high-temperature ruptures could not be ignored. There were critical flaws and errors in the safety review because it overlooked this possibility
3. Threat of a reactor core meltdown accident	
Plaintiff *local residents*	Although it was clear that the threat of a reactor core meltdown would have a major impact on local residents, by assuming that the frequency of such an occurrence was low, the safety review did not adopt strict criteria and denied its possibility. This was a critical flaw
Defendant *government*	As preventive measures were adequate, a reactor core meltdown accident could not be expected from a technical point of view. The claims by the local residents ignored completely all the preventive measures already taken

(continued)

Table 8.2 (continued)

Fukui district court's judgement	As the preventive measures were taken in the process of the safety review, the possibility of a core meltdown accident could not have been anticipated in the first place. Nothing in the evaluation of the risk of a core meltdown accident was unreasonable
High court's judgement	Fast-breeder reactors like that at Monju are still at the research-and-development stage and there is little operational experience. It is problematic to conclude that 'a reactor core meltdown accident cannot be anticipated from a technical point of view'. Obviously, the accident should have been anticipated

Source *The Mainichi Newspapers*, morning edition, 28 January 2003 (Translation by authors)

confirming the possibility of accidents to be undeniable.[9] Consequently, by stressing the flaws in the safety review process, they argued for the need to re-examine Japan's long-term energy policy as well.[10]

To summarise this section: Newspaper coverage of the Monju trial focused on the possible occurrence of low-probability events, the validity of the safety review, and Japan's long-term energy policy. However, what is crucial here is whether the High Court itself proactively judged the possibility of accidents on behalf of the experts and whether the courtroom, the currently used conflict resolution mechanism, is able to adequately handle the social conflicts arising from the construction and use of nuclear plants.

8.4 The Defendant's Response

8.4.1 The Government

Right after the High Court's decision, the Japanese government promptly made a rebuttal statement as the defendant. Table 8.3 describes the points that the government argued were problematic in the High Court's decision (METI 2003).

As is clear from Table 8.3, the government made a point of the invalidity confirmation, as well as the three issues reported by the newspapers. This was because the government intended to appeal to the Supreme Court. In Japan's three-tiered court system, cases that can be appealed to the Supreme Court are generally

---

[9] This tendency was apparent in the reporting of *The Mainichi Newspapers*, *The Nikkei*, and *Asahi Shimbun*. However, the High Court's finding on the possibility of an accident was a highly divisive issue. To say the least, experts did not fully agree on the possibility of this sort of accident occurring.

[10] Only *The Yomiuri Shimbun* dissented from the High Court's judgement on the possibility of accidents, and its editorial article also stated the ruling as 'dubious' (the morning edition of 28 January 2003).

Table 8.3 Arguments advanced by the government regarding the High Court's decision

1. About invalidity confirmation	
High court's judgement	Establishment of clear illegality is not required for invalidity confirmation of the construction approval of a nuclear reactor
Government's objections	It is a contravention of the precedents that demanded the establishment of material and clear illegality in order to invalidate the disposition
2. About the floor liner	
High court's judgement	There were critical flaws in the initial safety review because it failed to examine whether the thickness of a floor liner could adequately prevent the direct contact of sodium leaked from the secondary coolant system and concrete
Government's objections	In light of the established legal framework, matters such as the thickness of a floor liner and tubes were examined for their detailed design, and they have been adequately examined in conformity to the articles of the 'approval of the design and construction method'. Instalment of a floor liner was reviewed as a safety measure in the code of the 'instalment license' for the baseline design. There was no deficit in these processes
3. About simultaneous steam generator tube rupture accident	
High court's judgement	The occurrence possibility of a simultaneous steam generator tube rupture (high-temperature rupture) accident was undeniable. Failure to consider this possibility is a critical flaw
Government's objections	High-temperature rupture is a phenomenon in which leaked water from tube ruptures contacts with leaked sodium and results in simultaneous tube rupture at high temperature. This phenomenon was addressed fully by the instalment of manometers to detect pressure changes inside the steam generators, along with other preventive measures. As there is no possibility of this phenomenon occurring, there is no need to consider it in the review process. The High Court's judgement pointing out the possibility of a core meltdown accident assumed the possibility of high-temperature rupture. An accumulation of such unlikely assumptions is unrealistic
4. About a reactor core melt down accident	
High court's judgement	Even though the license applicant considered a reactor core meltdown accident as physically impossible, as long as the possibility was undeniable, the safety review process should have examined its possibility
Government's objections	From a technical point of view, it is inconceivable that the multiple layers of preventive measures would lose their functions simultaneously and that a reactor core meltdown accident would occur. Still, the safety review for the instalment of Monju did examine such possibilities, estimated the possible amount of energy generated with an adequate margin of safety, and confirmed the reactor's safety The High Court's decision demanded that the review process presuppose an amount of energy beyond rational thinking. It is irrational to combine this unlikely assumption with further unrealistic assumptions

Source METI (2003) (Translation by authors)

limited to violations of the Constitution, violations of legal interpretation, and contraventions of precedents. The making of factual mistakes does not qualify. Therefore, in seeking for invalidity confirmation, the government argued that the High Court's decision departed from the precedents of the past because it failed to establish the material and clear illegality of administrative disposal.

Administrative litigations have the characteristics of a lawsuit disputing the validity of procedures. The Monju trial centred on whether the construction permit by the government was issued in line with the code of rules stipulated by the Act on the Regulation of Nuclear Source Material, Nuclear Fuel Material, and Reactors (hereafter the Nuclear Reactor Regulation Law). The safety of the reactor itself was beyond the scope of this trial. In light of legal theory, then, the government's objection to the High Court's decision can be understood as reasonable because the decision contravened precedents by failing to establish the *material and clear illegality* of the disposal. However, in reality, it is very difficult to imagine that any administrative agencies would accept dispositions of such clear illegality. Thus, legal experts usually believe that as long as this requirement is imposed, plaintiffs do not stand much of a chance in administrative litigations. This is why it is argued that the pursuit of administrative litigations is pointless.

Regarding the judgements on low-probability events, unlike the newspapers' coverage, the government could not accept the High Court's decision as factual. In the appeal to the Supreme Court, then, while the government argued for contravention of the precedents, it also intended to refute the High Court's factual findings. For the government, it was not enough for the contravention of the precedents to be admitted; it had to seek the complete dismissal of the High Court's judgement, including the factual findings.

The next section reviews the government's objection to the High Court's factual findings on so-called low-probability events. The arguments put forward by the Nuclear Safety Commission (NSC), which was in charge of the safety review, demonstrated it most clearly.

8.4.2 The Nuclear Safety Commission

As a response to the High Court's decision, in March 2003, the NSC published a document titled 'Technical contentions on nuclear safety concerning the Nagoya High Court verdict at its Kanazawa branch on the prototype fast-breeder reactor Monju'. This document describes in detail the NSC's position on the technical contentions of the High Court decision. This section analyses the NSC's arguments described in the document, focusing on low-probability events (as the NSC does not officially publish an English version of this document, the translation in this chapter is by the authors).

8.4.2.1 Stepwise Regulation

At the beginning of the document, the NSC's position is explained. The role of the NSC is stipulated in the Nuclear Reactor Regulation Law. In this law, regulatory rules are prescribed by three steps: *design, construction*, and *operation* (Fig. 8.1).

For a prototype fast-breeder reactor such as Monju, the law prescribes that the minister of METI (when the installation license of Monju was applied, this was the prime minister) must hear, in advance, the opinion of the NSC. As Fig. 8.1 shows, the initial safety review of Monju by the NSC dealt with the baseline design only, and following the results of the review, the prime minister issued the construction permit. With regard to one of the disputed issues, *preventive measures against sodium leakage*, the NSC argued that specifications such as the thickness of a floor liner were matters of concern at the detailed design phase, not at the baseline design phase.

8.4.2.2 Basic Concepts of the Safety Review

The document, then, explains the basic concepts of the safety review. The NSC (2003) argues that, because it is impossible to ensure *absolute safety* when we utilise science and technology, the benefits of using science and technology must far exceed the risks of disasters. Furthermore, for the safety review of a nuclear reactor facility, because the potential hazards of accidents are much higher than those of other facilities are, the safety review must impose requirements that are more stringent. After stating these basic concepts of the safety review, the NSC explains its technical issues with the Monju High Court's decision.

The High Court's decision claimed that the issues reviewed by the NSC were not sufficient to ensure the safety of Monju, and the possibility of a severe accident was undeniable if the unintended events were to take place in a chain reaction. The NSC refutes this point as follows:

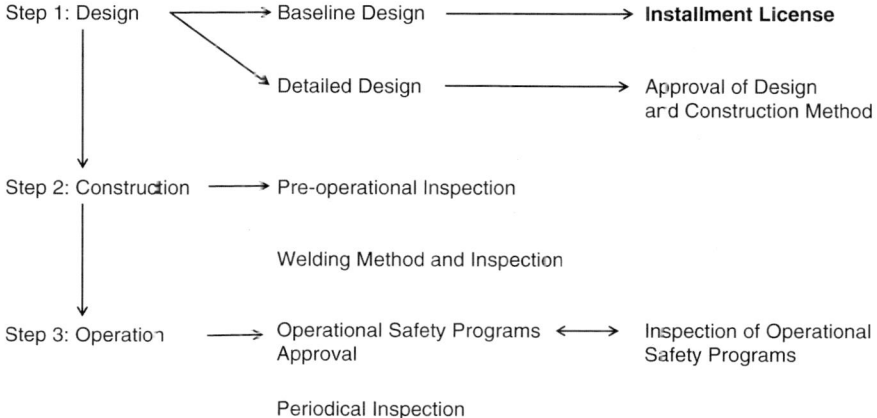

Fig. 8.1 Stepwise regulations. *Source* Komoda (2003)

In principle, by compiling a series of assumptions, it is possible to imagine severe accidents that could cause disasters of vast magnitudes, even up to infinity. And, *in theory, such possibilities cannot be excluded completely.*

The decision on how far the safety review process should examine such anomalous situations and events requires sophisticated expertise. In addition, engineering judgement conceives that there should be an upper limit of assumptions that need to be considered. This conception of a safety review is generally acknowledged worldwide in the science and technology communities (NSC 2003; italics by the authors).

Crucially, the NSC argues here that any judgement regarding the extent of possibilities that need to be considered, in other words, how many assumptions should be aggregated, should depend on engineering expertise. However, such engineering judgement is obviously not quantifiable and has a kind of tacit knowledge. This judgement, furthermore, may not always be precise but contain some margin of error, and there is no guarantee that all the experts in nuclear engineering and the atomic sciences will always reach consensus. We do not intend to criticise the use of engineering judgement itself here but would like to point out that the safety review process of public facilities, including nuclear facilities, which are potentially disastrous, inevitably involves some level of uncertainty that can be called 'engineering judgement'. What, then, should the judiciary system assess? We will come back to this question later.

8.4.2.3 Damage to a Floor Liner

Regarding the sodium leakage accident of December 1995, the NSC (2003) argues as below.

The safety review of the baseline design confirmed that a steel liner could prevent the direct contact of sodium and concrete. The validity of specifications, such as the thickness and structures of a floor liner, is a matter that demands a synthetic judgement that includes structural strength and workability. This judgement rests with the subsequent safety review for the detailed design.

However, at the trial, the NSC admitted that it was not aware of the possibility of floor damage caused by sodium leakage at the time of its safety review for the baseline design. The soundness of the safety review thus became a point of significant dispute.

8.4.2.4 New Findings after the Safety Review

Regarding the damage to a floor liner, the NSC (2003) states in the document as follows:

With respect to the corrosion of a steel floor liner, *the possibility of molten salt causing cause interfacial reactions was not widely known* (1) at the time of the safety review of the baseline design and was only observed later.

Although it was already known at the time of the safety review of the baseline design that *direct contact of steel and sodium in the presence of oxygen could cause an interfacial reaction to corrode steel* (2), *the experiments after the sodium leakage accident revealed for the first time that such corrosion could take place at a rapid rate* (3). Moreover, the rapid rate of corrosion that was observed for the first time in the experiment, so-called molten salt corrosion, was the result of infusing air intentionally to reaction points in order to facilitate the observation over the course of the experiment. It is possible that the oxygen in the infused air significantly promoted the reaction. Therefore, it clearly deviated from the conditions that should be considered in the safety review for the baseline design.

At the safety review of the JNC's application for the modification of the facility after the sodium leakage accident, the NSC re-examined, *just in case, the validity of the preventive measures of the baseline design regarding the installation of a floor liner for the secondary coolant*. As a result, the NSC *confirmed that, even if the molten sodium may cause corrosion, as had been discovered by the experiment after the accident, a floor liner did prevent direct contact between sodium and concrete. Thus, it was affirmed that there was no need to change the results of the safety review at the phase of the installation licence* (4) (numbers and italics are by authors).

The NSC's wording here has a very subtle nuance. First, it is not clear what was newly found after the accident. As far as reading this document, it seems that the NSC claims in sentence (3) that 'the rapid rate of corrosion', or 'molten salt corrosion', as the NSC refers to it, was newly discovered. Contrarily, right before this sentence, the NSC argues in sentence (2) that corrosion by interfacial reaction was already known about at the time of the safety review for the installation licence. Moreover, in sentence (2), the NSC states that 'interfacial reactions was not widely known about'.

Why did the NSC use such complicated expressions? We suspect the following reasons. The NSC's cardinal claim is that it had taken all possible measures to ensure the validity of the safety review. Therefore, if something were allowed to happen, the NSC would have to argue that the corrosion was *a phenomenon unknown at the time of the safety review*, and in this sense, that it was out of the NSC's control.

However, right after the accident, some experts argued that this kind of molten salt corrosion was commonly known about. For instance, Itaru Yasui, a professor at the University of Tokyo at the time, posted a comment on his webpage:

The High Court's decision on nullification of the Monju installation dated 28 January 2003 (Dialogue between a teacher and students).

Teacher: As for the holes in the steel liner, it seems that it really was unexpected for the NSC. Tatsuo Kondo, the former chairman of the NSC, said in *Asahi Shimbun*, 'it was not known at the time of the initial application for the installation in 1980 that leaked sodium would make holes in a steel liner'. However, *I suspect that this sort of corrosion was already common knowledge among researchers of thermodynamics in steelmaking and refining*.

Student: At the time of the accident, researchers who make the phase diagrams of molten salt of oxides said that it was common sens (Yasui 2003; Translation and italics are by authors).

If the NSC really were unaware of this possibility, then the validity of the safety review would be in doubt. The NSC must have wanted to avoid this situation.

Therefore, it seems that the NSC employed the following logic: although the possibility of corrosion 'was not widely known' (sentence 1), the NSC was already aware of this in general (sentence 2), but the details were only revealed for the first time by the experiments after the accident (sentence 3).[11]

On the basis of this logic, it would seem there was no flaw in the NSC's safety review system. At the same time, such argumentation by the NSC also elucidates the fact that it is very difficult to review the safety of large-scale complicated systems such as nuclear-related technology because it demands a wide range of expertise. The difficulty in consolidating the best available expertise appropriately, for some situations, may cause a serious problem.

After presenting a complex description of the *new findings*, the NSC comments in sentence (4) on the relationship between these findings and the safety review. Its logic is that the damage to floor's steel liner by the sodium leakage accident suggested the possibility of corrosion and then the PNC performed the experiments as a comprehensive safety review of Monju to improve safety. As a result, the experiments found that even if the molten salt caused corrosion, the steel liner would prevent direct contact of sodium and concrete. This demonstrates that there was no flaw in the initial safety review for the installation permit.

The NSC's logic sounds like hindsight and prompts a couple of questions: 'Is there any scientific knowledge that remains undiscovered?' and 'Are there possibilities that such undiscovered knowledge might have a serious impact on safety?' In a sense, these questions may be impossible to answer. The NSC would want to respond that there is no way of discussing these issues until new discoveries are made. However, at least one piece of new knowledge was discovered, and it provides the rationale to raise these questions. The questions should be distinguished from the common accusation that the public demands absolute safety with no risk. Rather, having given up the demand for absolute safety, it is more important to consider the safety of a technological system that inevitably entails unknown risks.

After stating its views on the new findings, the NSC refers to the objectivity of the safety review. 'As the safety review requires a high quality of expertise, the NSC set a screening panel with *the highest level of experts* in relevant disciplines' (italics are by authors). In order to assure objectivity, the NSC 'selects experts carefully,' 'requests a third party to perform calculations and experiments for verification, and to compare evaluation results in other facilities of the same kind'. In addition, 'the NSC considers that specific evaluation methods should also be decided on the basis

---

[11] The High Court's ruling stated that, 'The appellate acknowledged that neither staffs of PNC nor the members of NSC who were involved in the safety review for the construction permit possessed knowledge about the corrosion mechanism of high temperature sodium and steel. (Meanwhile, as the NSC's accident investigation working group pointed out, they could have accessed this knowledge at the time of construction permit if they had been aware of the issue.)'. The Nuclear Industrial Safety Agency (NISA) rejected this argument, maintaining that, 'The knowledge that the Electrochemical Society of Japan, to which the NSC's working group commissioned the investigation, had used the phrase "could have accessed" is knowledge that "corrosion would occur." It did not imply that there was quantitative data to prove corrosion, nor that corrosion can make holes in a steel liner of 6 mm thick within a few hours' (NISA 2003).

of expert opinion as well'. The NSC also maintains that 'the safety review at the installation of Monju was performed on the basis of *engineering judgement* using the best available expert knowledge at the time, and the review sufficiently examined the baseline design in terms of safety' (italics are by authors).

In short, the NSC argues that it performed the safety review of Monju for the installation permit using *engineering judgement with the best available expert knowledge from the highest level of experts*. When such expert judgement lies at the centre of a conflict, then, what should the judicial system be examining in the trial?

8.5 The Responses of Legal Experts

As we mentioned earlier, administrative litigations on nuclear-related facilities must refer to the Supreme Court decisions of 2005 regarding the Ikata Unit 1 and the Fukushima II Unit 1 nuclear power plants. In the Monju trial, when the defendant appealed the High Court's decision to the Supreme Court, the defendant also referred to these decisions from 1992, maintaining that the High Court's decision on Monju departed from these precedents.

8.5.1 *The Supreme Court's Decisions on the Ikata and the Fukushima II Nuclear Power Plants Trials*

The Ikata Nuclear Power Plant (hereafter the Ikata NPP) is located in Ehime and is owned by the Shikoku Electric Power Company. The company applied for installation of the Ikata NPP in 1972 and the prime minister approved the installation. In 1973, 33 local residents filed an administrative lawsuit to cancel the approval. The plaintiffs' request was dismissed in the first and second instances, and finally in the Supreme Court in 1992. A variety of issues were disputed, but the main points concerned four subjects:

1. Whether the government's procedures to approve the installation were in violation of Article 31 of the Constitution, which guarantees due process of law, including public involvement and disclosure of the relevant information.
2. Whether the Nuclear Reactor Regulation Law, having no detailed and specific criteria for the safety review, was in violation of Articles 31 and 41 of the Constitution. Article 41 says, 'The Diet shall be the highest organ of state power, and shall be the sole law-making organ of the State' and guarantees separation of powers.
3. Whether the government's administrative discretion was allowed in the installation of a nuclear reactor, and how that discretion should be tested legally.
4. What a safety review for the installation of a nuclear reactor should examine (Takagi 2004, p. 194).

As regards points (1) and (2), the decisions rejected the plaintiffs' claims and most legal experts regarded the decisions as legitimate interpretations of the Constitution (Takagi 2004, p. 196). However, some jurists remarked that these points of issue should not be judged solely on an interpretation of the Constitution, especially with regard to the obligation to involve in the decision-making process those who may be affected by the administrative disposal. This should be further examined from the standpoint of STS as well. As for the points of (3) and (4), the Supreme Court decision is summarised below.

8.5.1.1 Approval of Stepwise Regulation to Ensure Nuclear Safety: Baseline Design and Detailed Design

The Supreme Court decision on the Fukushima II nuclear power plant (hereafter the Fukushima II NPP) approved stepwise regulation to ensure the safety of the nuclear reactor. This meant that the scope of the safety review for installation of a nuclear reactor was limited to the baseline design only. It did not include detailed matters, such as methods for final disposal of nuclear waste, reprocessing and transportation of the spent nuclear fuel, reactor decommissioning, man-machine interface, and preventive measures for stress corrosion cracking (SCC).

This decision is problematic in two respects. First, as long as administrative litigation remains within the frameworks of the Atomic Energy Basic Act and the Nuclear Reactor Regulation Law, social aspects of nuclear power plants and other matters that determine the shape of our civilisation are out of scope of the judicial review (Takahashi 1993, p. 51). Where, then, might we be able to discuss the direction of our civilisation? The decision thus highlights the shortcomings of our society in terms of its capacity to settle conflicts, because filing a lawsuit is currently the only way for citizens to express their views regarding nuclear facilities.

Second, the Supreme Court's decision on the Fukushima II NPP held that the examination of the man-machine interface[12] was not within the scope of the safety review for the baseline design. The legitimacy of this judgement requires examination, ultimately, not in terms of the interpretation of the law but in terms of design engineering knowledge. In short, it demands *engineering judgement*, and as we mentioned earlier, there is no guarantee that experts' opinions will always reach consensus. In the first place, the man-machine interface should be of concern throughout the entire process—baseline design, detailed design, and management of operations. Thus, it may be fundamentally unfit for stepwise regulation. This also illuminates a limit of administrative litigation, because it mainly adheres to the interpretation of laws (Abe et al. 1993, p. 11).

---

[12] Man-machine interface was an issue raised by the Three Mile Island and the Chernobyl nuclear accidents, which required special attention to be paid to the safety management of nuclear facilities. In this sense, man-machine interface was also *new knowledge*, just like the corrosion of a steel liner (Takahashi 1998, p. 201).

8.5.1.2 The Validity of Disposal Examined in Light of the Best Available Scientific Knowledge

The Supreme Court's decision on the Ikata NPP held that, according to stepwise regulations, the safety review process for the baseline design examined, while presupposing that a nuclear reactor would be built exactly as planned in the baseline design, whether or not sufficient preventive measures were taken to prevent accidents when the plant began operation.[13] Then, while taking into account that constant progress was being made in science and technology, the Supreme Court judged whether the approval for installation violated the law in light of the current level of knowledge in science and technology, even if the safety review was valid at the time of installation. If the examination revealed any inadequacies in the safety measures, then the court would cancel the installation permit because the disposal would be proven illegal at the time of the judicial review.

Toshifumi Takahashi, a research law clerk for the Supreme Court, explained the term 'current level of knowledge in science and technology' in the following way:

> Although accident prevention measures included in the baseline design were considered adequate at the time of the safety review for installation approval, if these measures are inadequate in light of the current level of widely accepted knowledge, and if there is a high possibility of severe accidents when a nuclear reactor is installed as planned in the baseline design and it starts operation subsequently, then the installation permit presupposing the safety of the nuclear facility should be cancelled (Takahashi 1993, p. 57, translation by authors.).

In this sense, administrative litigation concerning nuclear facilities is different from that concerning medical malpractice, which examines whether or not there was negligence in light of the medical standards at the time of the medical accident.

However, some jurists said they were uncomfortable with this logic, because even if the safety review was carried out while based on the best available knowledge, the installation permit could become illegal afterwards. Some would argue that even if new knowledge found flaws in the safety systems of the baseline design rather than the earlier decision being treated as retroactively illegal, the court should consider the earlier decision as still legally valid but should order the administrative agency to address the flaws discovered by the new knowledge promptly (or cancel or suspend the disposal). Should the agency neglect this obligation, then citizens would be allowed to file a lawsuit to request due diligence (Takagi 2003, 2004; Nakagawa 2004). In order to adopt such a process, however, the Administrative Case Litigation Act would have to be revised. Here, the interface between the judicial system and legislation becomes an important issue.

---

[13] As this judgement indicates, some legal experts note that this kind of litigation has the characteristic of putting the future on trial, because the resulting judgment would have a major influence in shaping the future society. This make sense: a lawsuit about a nuclear power plant must examine the safety of the equipment to be used for many years, as well as its possible negative impact, long before any harm may appear (Sato 1993, p. 39).

8.5.1.3 The Administrative Judgement in Respect of Matters Requiring Expert Knowledge

The Supreme Court decision on the Ikata NPP trial stated that the decision on whether or not the safety review of the nuclear power plant conformed to the Nuclear Reactor Regulation Law was entrusted to the prime minister's rational judgement, while a court examined whether or not there had been anything unreasonable about the process of administrative disposal.

This said, the court examined, in light of the current levels of science and technology, the following:

1. Whether the criteria actually used for the safety review included any unreasonable aspects.
2. Whether there were any errors or deficits that could not be overlooked in the screening process of the Atomic Energy Commission and the Nuclear Reactor Safety Specialty Committee of the NSC, which had concluded that the baseline design for the Ikata NPP met the required criteria.

If the court found anything unreasonable or any errors or deficits that could not be overlooked and the administrative disposal was made on the basis of such a flawed review and process, then, the resulting disposal should be also considered unreasonable (Takahashi 1993, p. 56).[14]

This logic illuminates the fact that, in order to examine the validity of the safety review for the installation approval of a nuclear reactor in an administrative litigation, a court focuses on rationality and adequacy of the safety review criteria and process, not the safety of the nuclear reactor itself. This is why an administrative lawsuit is called a 'lawsuit of procedures'. Hikaru Takagi remarked in this regard that:

> Civil lawsuits, in which residents face off against the licensee, not the government, examine the safety of the nuclear reactor itself. Consequently, the judges, who are inexpert in engineering, are expected to assess the safety of a reactor and to make determinations. By contrast, administrative lawsuits examine whether or not the government and the minister took appropriate lawful procedures to issue the installation approval. In this case, the court respects the 'discretionary powers' of administrative agencies (Takagi 2004, translation by authors).

---

[14] While the lawsuit on the Ikata NPP was a suit for cancellation of a construction permit, the lawsuit on Monju was one for invalidity confirmation of a permit. The requirement to establish 'any unreasonableness, errors, or deficits that could not be overlooked' in the process of screening and making a judgment is in pursuit of cancellation. The generally accepted requirement for invalidity confirmation, by contrast, is to establish the *material and clear illegality* of administrative disposal. How these different requirements relate to each other remains open to dispute. Some have argued that the High Court's ruling on Monju equated these requirements when they should be distinguished (Takagi 2003).

The Supreme Court decision on the Ikata NPP exemplifies that, in administrative litigation, the 'discretion'[15] of administrative agencies is greatly respected.

8.5.2 The Logic of the Administrative Lawsuit

If we consider the Supreme Court's decisions on the Ikata NPP and the Fukushima II NPP trials as a benchmark, they demonstrate how astonishing the High Court's decision in the Monju trial was. It was astonishing because, as in the Ikata NPP and the Fukushima II NPP trials, an administrative litigation regarding the safety review for installation of a nuclear power plant usually presupposes that (1) stepwise regulation is regarded as legitimate; (2) the court examines whether or not there were any legal violations in administrative procedures, not the safety of the reactor itself; and (3) experts' discretion is generally acknowledged. The High Court's decision in the Monju trial questioned these premises. Although many legal experts hold the decision in high regard for its demand that only material illegality need be established—rather than *material and clear* illegality—to request the invalidity confirmation of administrative disposal, there are many jurists who question the legitimacy of this decision from the perspective of administrative litigation as a lawsuit that examines the lawfulness of administrative procedures.

In administrative litigation on science and technology, the role of judicial system is understood in two ways: while some jurists believe that the judicial system should make determinations on scientific matters, other jurists think that the judicial system should make determinations solely on procedural aspects. Between these two extremes, a wide variety of views have been expressed (Takagi 2003; Takahashi 1998).

8.5.2.1 A Theory of Substituting Substantial Judgement (Judging Substantively on Behalf of the Government)

One way to think about the role of administrative litigation is that a court examines the legitimacy of the disposal from the same position as an administrative agency, compares it with other administrative disposals, and makes determinations. For a nuclear reactor, the court examines specific matters (e.g. the safety of the anti-rupture system of a pressure vessel) while hearing from expert witnesses.

Many aspects of this view have been seen as problematic, and this way of thinking is not considered to be mainstream in academic circles. Aspects that have

---

[15] The High Court's judgement did not use the term *discretion*. This is because, in the judicial process, the term *discretion* generally means the discretionary powers of making judgements on political and policy matters. Takahashi (1993, p. 55) argues, however, that the key point of the Supreme Court ruling on the Ikata NPP was that this ruling admitted the discretionary powers of administrative agency on *specialized and technical matters*.

been criticised include (1) the fact that this approach requires the judiciary to serve as the second process of administrative approval and (2) that it turns an administrative lawsuit into a trial by expert witnesses. In particular, many legal experts have expressed their concerns that a judge who is inexpert in science and technology makes independent determinations by hearing from expert witnesses. Such concerns become more compelling when expert witnesses have differing opinions. For instance, when expert witnesses express different opinions regarding the validity of an emergency core cooling system (ECCS), some arguing that 'the validity has been substantiated by experiments of each part and computer simulations', while others argue that, 'such data is not sufficient to prove the validity, and more scale model tests must be done', then whether or not the judge is able to be assess these testimonies becomes critical.[16]

Because of these concerns, some jurists assert that, for issues requiring expert knowledge in science and technology, a court should respect the discretionary power of an administrative agency. In other words, a court should not directly make determinations about substantive matters regarding science and technology but should engage indirectly with them in some way.

8.5.2.2 A Theory of Procedural Examination of Substantiality

The opposing view on the role of administrative litigation is that a court should examine whether anything was unreasonable in the process of the administrative authorities' review and approval. If any unreasonable points are found, then the approval should be also considered unreasonable and thus illegal, and so the disposition should be cancelled. This view currently predominates among academics. The Supreme Court decision on the Ikata NPP is regarded as adopting this position (Abe et al. 1993).

Going back to the High Court's decision in the Monju trial, Hikaru Takagi argues that this decision constituted a 'substituting substantial judgement',[17] although this was not clear. The High Court's decision states that:

> The judicial review does not need to find proactively the substantial risk of the above event (the release of radioactive materials stored in the containment vessel into the environment), but should examine only whether or not the substantial risk is deniable (1). In an appeal lawsuit against the installation approval of a nuclear reactor where the legitimacy of the judgement regarding the safety of a nuclear facility is disputed, the judicial review and decision should be made on the basis of whether there was anything unreasonable about the judgement of the administrative authorities (as in the Supreme Court's decision on the Ikata NPP), because a court is not in a position to judge on behalf of the administrative agency (2). (Nagoya High Court Kanazawa Branch 2003, translation and numbers are by authors).

---

[16] Alvin Weinberg (1972) referred to such questions as *trans-scientific* questions.

[17] This means that the judge makes a decision rather than an expert or administrative agency.

Thus, this decision clearly followed the Supreme Court decision on the Ikata NPP and did not adopt the approach of substituting substantial judgement. However, Takagi argues that:

> The first sentence of this decision (1) showed its weakest point. Even if the probability of *the worst situation* was considerably low, it was still not zero. Thus, in order to examine whether or not there was *substantial possibility*, a court needed to make a certain determination. If a court *proactively* finds this possibility, then, it would be a substitution of substantial judgement. As a court must maintain the premise specified in the second sentence (2), it takes a position that a court should examine 'whether the substantial possibility is deniable or not'. However, as long as *substantial possibility* means a certain level of probability, a court cannot reach a determination without substituting administrative judgement (Takagi 2003, translation by authors).[18]

As Takagi argues, this decision can be understood as having examined the possibility of a certain event and having made a *substantial judgement* differing from that of the defendant (experts). In other words, the point is whether a very low probability should be regarded as a substantial possibility or as *not going to happen from the technological point of view*. For the experts in nuclear engineering, this is an *engineering judgement*, which is different from a *scientific fact*. For a judicial review, the question is whether this *engineering judgement* should be included in the scope of the experts' discretion or not.

Takagi also referred to the review in the final appellate instance, the Supreme Court. If the High Court's decision was covertly *substituting substantial judgement*, then the Supreme Court should return to the conventional approach to judicial review of examining the validity of procedures, as in the decision on the Ikatau NPP. The point of issue would then be whether the High Court substituted substantial judgement, not whether there were errors in the defendant's judgement about low-probability events. Moreover, in principle, the Supreme Court should not examine the factual findings of the lower court. Therefore, if the High Court's *substantial judgement* concerned factual findings, then the Supreme Court is not allowed to examine its validity in terms of its factual findings. As the defendant, the government would not be happy with this principle because it would want the Supreme Court to examine the validity of the High Court's judgement on low-probability events.

8.5.2.3 A Theory of Controlling the Code of Conduct

As Takagi remarked, it is a subjective decision to choose which experts' engineering judgement to trust and what aspects of the judgement should be considered. A third way of thinking about the role of administrative litigation is that the judiciary should not make any judgement about engineering at all. This view, in a sense, is a refinement of the theory of *procedural examination of substantiality*, as

---

[18] Takagi (2003), in quoting the judgement, changed the term in the second sentence from 'substantial risk' to 'substantial possibility'.

in the decision on the Ikata NPP. In this view, administrative agencies have only to do their best at each step of regulation (Takagi 2003, p. 29).

This view seems very clear as a matter of logic. However, it can lead to a situation in which, if the administrative agency conducted the safety review in compliance with the Nuclear Regulation Law (including experts' investigations required by the law), a court may need to accept the disposal almost completely. Under these circumstances, the discretion of the experts in the administrative agency is nearly total, and conflicting views between experts for the defendant and the plaintiff would not affect the decision at all. As a result, the court would examine only how new findings in science and technology should be treated.

8.5.3 Limitation of Administrative Litigation

8.5.3.1 Low-Probability Events and Relative Safety

In summary, on the one hand, it is difficult for the judiciary to make a substantial judgement in administrative litigation. First, it turns an administrative lawsuit into the second procedure for administrative approval. Second, it turns the lawsuit into a trial by expert witnesses. On the other hand, if a court adheres to procedural examination only, although it seems rational in light of legal theories, it may leave the discretion of the experts in the administrative agency almost entirely unquestioned, which would undoubtedly raise public concerns.

In lawsuits on nuclear power plants, the experts' discretion on low-probability events is frequently discussed within a framework called *relative safety*. This framework presupposes that safety is a matter of degree and that it is impossible to ensure absolute safety for engineering facilities. Thus, safety should be examined in relation to social utility, and should be accepted if the dangers do not exceed a socially acceptable level. Toshifumi Takahashi, a research law clerk for the Supreme Court, remarks:

> When an administrative agency reviews the safety of a nuclear reactor for installation approval, it must decide on certain criteria to judge the safety and to approve installation. In other words, the agency must determine how safe is safe enough, and, for this purpose the agency must consider what society regards as an acceptable level of risk. Under this situation, it is hard to deny that this decision must rely on the discretion of experts in the administrative agency who are in charge of nuclear safety. (Takahashi 1993, p. 55, translation by authors)

This view is sound in general terms. Takagi also maintains that the Nuclear Regulation Law should be interpreted in line with this view. Besides, in setting a range of criteria, the NSC included low-probability events in their considerations, even if such events are regarded as 'never going to happen from a technical point of view'. Thus, experts' scrutiny should be understood as being made in conformity with a principle of *relative safety*. Takagi criticised the High Court's decision on Monju, which held that 'the safety review should examine low-probability events because they could happen'. Takagi argued that because this denies the NSC's execution of experts' discretion, the decision covertly fell into *the substitution of substantial judgement*.

What we would like to ask is why the judgement of the experts from the administrative agency is the only one allowed under *experts' discretion*. When trying to refine legal theories of administrative litigation, jurists have not sufficiently considered the problem of conflicting views among experts. Why should the judgement on relative safety rest with the *discretion of experts* from the administrative agency? There is no doubt that this question will never be raised in administrative litigation.

8.5.3.2 The Struggles of Legal Experts

At present, for administrative lawsuits, it seems that most legal experts adhere to a theory of procedural examination, if not a theory of controlling the code of conduct, and reject a theory of substituting substantial judgement. This looks logically correct in light of legal theories. However, jurists themselves appear to be struggling to balance competing goals in the pursuit of both logical correctness and resolving social conflicts.

Such jurists' struggles are evident in the records of the discussion among the prominent scholars of administrative jurisprudence published in a journal targeted at legal professionals in 1992. For instance, in the discussion, it was proposed that even presupposing the principal role of an administrative lawsuit to be procedural examination, the rules for examination should be clarified further.

Hisashi Kouketsu argued that while the law should provide specific rules regarding matters that may potentially have a high impact upon the life and health of citizens, in light of the abilities of legislators and the rate of advancement of science and technology, it is difficult to say that the current system, in which the Diet legislates general laws and the administration decides specific matters, is unconstitutional.

> At least in light of legislative theories, rules should be written as specifically as possible in Cabinet Orders and Ministerial Ordinances. The process of making these rules should be, a little like in the US, open to the public, including people who are concerned about the safety of a nuclear facility. (Abe et al. 1993, p. 12, translation by authors)

Regarding experts' discretion, Takehisa Awaji raised the issue of who should control the safety review and how, remarking:

> Many issues can be disputed, such as the ways to choose committee members, to conduct reviews, to collect data to support the review, and so on. The main challenge is to find a sound procedure to control the safety review and to avoid any mistrust. (Abe et al. 1993, p. 14, translation by authors).

Yasutaka Abe pointed out that experts' discretion might lead to the situation that he called 'the pathology of professional organisation':

> It is only after the result of the safety review is examined by experts of differing opinions, and the judgements by the administrative authorities are proven to be more rational than those of others are, that experts' discretion can be respected (Abe et al. 1993, p. 17, translation by authors).

Referring to conflicting views between the experts for the plaintiff and those for the defendant, Kouketsu maintained that:

> There is another issue with what can be regarded as experts' discretion. Concerning the matter of the emergency core cooling system (ECCS), for instance, no matter how much the experts for the defendant argue in its favour, the experts for the plaintiff will never agree definitively, saying that it depends on what is regarded as science. However, once a court judges that the ECCS is a matter that depends on the discretion of the experts of the administrative authorities, then there will be a determination (Abe et al. 1993, p. 17, translation by authors).

Without doubt, all of the above points are thoroughly sound. Who should control the safety review? How should rules be made? Is it necessary to involve people who may be affected? Can we trust the experts' discretion? How should conflicting views among experts be handled? Why are only the judgements of experts from the administrative authorities allowed discretion? These questions should be of great significance but within the framework of an administrative lawsuit, they will never be raised.

A last remark on this discussion. Awaji maintained that:

> After all, it (the current safety review system) is a process in which the potential casualty of an accident has no channel for expressing their voice. In this system, the premises are that all of the specific matters should be decided within the experts' discretion; that the siting process does not require the involvement of the potential victims; that legislation should decide matters such as whether or not to hear from those concerned, whether to involve them in decision making, and whether to disclose relevant information. The legitimacy of such a system should be questioned (Abe et al. 1993, p. 30, translation by authors).

This discussion demonstrates that there is a discrepancy between the reach of legal scholars who specialise in administrative litigation and the increase in the range of practical issues brought into court. The discussants expressed their concerns about the current situation, in which all of the issues they regard as crucial but beyond their reach are labelled as legislative matters.

In any case, within an administrative law framework, it is difficult to examine the safety of a nuclear power plant in court, both in theory and in practice. Again, this highlights the extraordinariness of the High Court's decision in the Monju trial. In other words, the High Court examined issues that the Diet should have pondered, and they did so within a framework of administrative litigation that, at least for some people, seems to be no longer adequate for judging some issues in modern society, such as the safety of NPPs. Consequently, the High Court pronounced a judgement that seemed somewhat unreasonable in light of the current understanding of legal theories (Sanbe 1993; Sudou 2003; Hokimoto 2003; Yamashita 2003; Aizawa et al. 2003). This point should be elaborated on, beyond the Supreme Court's judgement on Monju.[19]

8.6 Conclusion

In summary, the analysis of the High Court's decision on Monju has revealed the limitations of administrative lawsuits regarding nuclear power plants to be several:

1. Administrative litigation concerning a nuclear facility examines whether or not the NSC conducted the safety review in compliance with the relevant laws, such as Nuclear Reactor Regulation Law. The safety of the reactor itself is beyond the scope of the judicial review.
2. Administrative litigation on a nuclear facility does not make a judgement about the safety of the nuclear reactor itself.
3. As for the safety of a nuclear power plant, in principle, the discretion of experts from the administrative authorities is treated with great respect.
4. In administrative litigation, stepwise regulation of installation and operation of a nuclear power plant is taken as a premise, and matters such as the final disposal of nuclear waste and methods of reprocessing spent nuclear fuel are beyond the scope of a judicial review, let alone any social issues relating to our civilisation, such as what a society that uses nuclear energy should be like.

---

[19] On 30 May 2005, the First Petty Bench of the Supreme Court quashed the High Court decision. Thus, the judgement of the Fukui District Court, which quashed the request for nullification of the construction permit of Monju, became final. The Supreme Court decision did not explicitly mention that the High Court required that only *material* illegality need be established for the invalidity confirmation of the construction permit of Monju, and not *clear* illegality. Instead, the Supreme Court judgement endorsed the judgement of the government agencies regarding 'the secondary coolant leakage accident' (the so-called sodium-leakage accident), 'breakage in a heat transfer tube', and 'event of unprotected loss of flow in the primary cooling system', and determined that the High Court judgement contained a violation of interpretation of the Nuclear Reactor Regulation Law. In other words, the Supreme Court's judgement granted the government agencies wide discretionary powers, including judgement over what matters should be included in the scope of the *baseline design* in the step-wise regulations stipulated by the Nuclear Reactor Regulation Law (Supreme Court of Japan 2005). This judgement was in line with the Supreme Court's decisions on the Ikata NPP and the Fukushima II NPP. In terms of the legal theories regarding administrative litigation, this judgement is similar to *the theory of controlling the code of conduct* that we mentioned in this chapter. However, this judgement does not mean that a judicial court confirmed the safety of Monju; it eliminated any concerns over the operation of Monju even less. In addition, while from the legal perspective, resumption of operations at Monju became legally permitted, from an engineering perspective, the resumption at Monju, which had stopped operations for 10 years, would have involved considerable risk. This is because it was not supposed to stop operations for 10 years when it was designed.

However, from the standpoint of STS, the analysis also revealed that there are some problematic issues in using an administrative litigation framework as the main mechanism for resolving social conflicts concerning nuclear power plants:

1. As the issue of new findings, such as the *corrosion of a floor's steel liner*, demonstrates, in cases of consolidated and mega engineering systems such as nuclear power plants, the relevant forms of expertise are diverse, and it is not easy to assemble and utilise such expertise appropriately.
2. There have been no discussions at all about *engineering judgement*, including what it implies, what its theoretical status is, and how to handle it, particularly in consideration of low-probability events.
3. Judgement on what should be included in the *baseline design* and what should be in the *detailed design*, as well as judgement on how safe is safe enough, are entirely at the discretion of the administrative authorities' experts.
4. When experts' views are disputed, the views of the administrative agencies' experts are predominantly favoured.
5. The question of the appropriate venue for the discussion of nuclear power plant safety and the social aspects of nuclear facilities remains unanswered. As a result, these issues are usually brought up in administrative litigation. This situation highlights the woeful lack of competence to resolve social conflicts concerning nuclear power plants in Japanese society.
6. There is no mechanism for local residents, as a concerned party, to participate in the process of making an administrative disposal.
7. In academic circles, too, although legal experts are aware of problems with the current system of administrative lawsuits, they ascribe all the shortfalls to matters of legislation and are yet to propose any changes to improve the judicial system.

What lessons can we learn from this analysis? First, we should not expect too much in the way of administrative litigation regarding installation permits for nuclear power plants. In this regard, we also found that newspaper reports were somewhat misleading. Newspapers reported the story as though the High Court had confirmed the possibility of low-probability events' occurrence (as we have seen, the decision featured 'a covert substituting substantial judgement'), and as if a court had made a judgement on the safety of the reactor. In addition, the tone of the newspaper reports expected the court to make such judgements. If the media coverage promoted extravagant expectations from the administrative lawsuit among citizens, then it would bitterly disappoint them afterward, as indeed the Supreme Court decision in the Monju trial did.

Perhaps we need to discuss how we can design and construct an appropriate system for deliberations on the social aspects of nuclear energy use, including the future direction of our civilisation. In addition, although the Supreme Court's decision on the Ikata NPP rejected it, we also need to make rules that require the involvement of local residents in siting decisions, no matter whether or not it refers to Article 31 of the Constitution.

The biggest obstacle in moving forward in this way is the issue of experts' discretion. In particular, we need to re-examine the extent to which we should depend upon the discretion of a limited number of experts with 'engineering judgement', regarding the issues of low-probability events, distinctions in baseline and detailed design, and the concept of relative safety.

References

Abe, Y., Awaji, T., Koketsu, H., Kobayakawa M., & Takahashi, S. (1993). Zadankai: Ikata Fukushima dai-ni genpatsu soshyou saikousai hanketsu wo megutte (Round-table talk on Ikata & Fukushima No. 2 NPP lawsuits). *Jurist, 1017*, 9–35. Retrieved April 15, 2014 from http://www.yuhikaku.co.jp/static_files/shinsai/jurist/J1017009.pdf.

Aizawa, I., Akimoto, Y., Iwabuchi, M., Kondo, S., Fukushima, M., Murakami, Y., & Yamaji, K. (2003). Monju hanketsu koredake-wa itteokitai (Comments on the Monju Decision). Energy Forum, *49*(580), 48–55.

CNIC, Citizens' Nuclear Information Center. (2003). Major victory to blow nuclear fuel cycle policy: The ground-breaking decision on the Monju fast breeder reactor. *Nuke Info Tokyo, 93* (Jan/Feb), 4. Retrieved April 15, 2014 from http://www.cnic.jp/english/newsletter/pdffiles/nit93.pdf.

Fukui Prefecture Citizens against Nuclear Power. (1996). *Kousoku-zoushyoku-ro no Kyofu (Fear of fast breeder reactor)*. Tokyo: Ryokufu Shuppan.

Hokimoto, I. (2003). Monju kousoshin hanketsu to kongo no genshiryoku gyousei (Appeal court decision on Monju and NPP policy). *Houritstsu Jihyou, 75*(4), 1–4.

Jasanoff, S. (1995) *Science at the bar: Law, science, and technology in America*. Cambridge: Harvard University Press.

Komoda, Y. (2003). Gensiryoku-hatsuden to kaku-nenryou-saikuru no anzensei (Safety of nuclear power plants and nuclear fuel cycle). Paper presented at the panel discussion on 'Safety of Nuclear Power Plant and Nuclear Fuel Cycle' at the Institute of Electrical Engineers of Japan, Tokyo.

METI, Ministry of Economy, Trade, and Industry of Japan. (2003). Kousoku-zoushyoku-ro Monju gyousei-soshyou no gaiyou (Outline of Monju administrative lawsuit). http://www.hiroi.iii.u-tokyo.ac.jp/index-genzai_no_sigoto-gensiryoku-monju-sosho-04.pdf.

Nakagawa, T. (2004). Monju jiken sashimodoshii-go kousoshin hanketsu (The high court decision on Monju). *Kankyohou Hanrei Hyakusen Bessatsu Jurist, 171*, 202–203.

NISA, Nuclear Industrial Safety Agency. (2003). Monju koutousaibansho hanketu no gaiyou to mondaiten (Outline of and problems with the decision in the Monju lawsuit). http://www.meti.go.jp/committee/downloadfiles/g30314a10j.pdf.

NSC, Nuclear Safety Commission. (2003). Kousoku-zoushycku-ro 'Monju' ni kansuru Nagoya koutou-saibansho Kanazawa-shibu no hanketsu ni kakawaru genshiryoku anzen no gijyutsu-teki ronten ni tsuite (Technical contentions on nuclear safety concerning the Nagoya High Court verdict at its Kanazawa branch on the prototype fast-breeder reactor at Monju). Retrieved April 15, 2014 from http://www.nsr.go.jp/archive/nsc/anzen/sonota/kettei/20030326.pdf.

Sanbe, N. (1993). Genshiryoku to hou no kongo no kakawarikata (Atomic energy and law in the future). *Jurist, 1017*, 36–42. Retrieved April 15, 2014 from http://www.yuhikaku.co.jp/static_files/shinsai/jurist/J1017043.pdf.

Sato, H. (1993). Ikata Fukushima dai-ni genpatsu soshyou saikousai hanketsu no ronten (Issues concerning the supreme court's decision of Ikata and Fukushima No. 2 nuclear power plants). *Jurist, 1017*, 36–42. Retrieved April 15, 2014 from http://www.yuhikaku.co.jp/static_files/shinsai/jurist/J1017036.pdf.

Sudou, S. (2003). Monju genpatsu kousoshin hanketsu (Appeal court's decision on Monju fast breeder power plant). *Gekkan Hougaku Kyoushitsu, 271*, 44–49.
Supreme Court of Japan. (2005). Judgment concerning what matters are included in the scope of matters concerning safety of the basic design of a nuclear plant that should be subjected to the safety examination for granting permits. Case No. 2003 (Gyo-Hi) No. 108. Dated 30 May, 2005. Retrieved April 15, 2014 from http://www.courts.go.jp/english/judgments/text/2005.05.30-2003.-Gyo-Hi-.No.108.html.
Takagi, H. (2003). Sairyou tousei to mukou (Discretionary control and nullity). *Jichikenkyu, 79*(7), 41–58 and *79*(8), 23–42.
Takagi, H. (2004a). Ikata genpatsu jiken: kagaku mondai no sihou-shinsa (Ikata nuclear power plant lawsuit: Judicial review of scientific issues). *Kankyohou Hanrei Hyakusen Bessatsu Jurist, 171*, 194.
Takagi, H. (2004b). Monju soshyou no kyoukun (Lessons learned from the Monju trial). Paper presented at Shimin-shyakai ni okeru kagaku to shihou wo kangaeru shimpojium (Symposium on Science and Judiciary in Civil Society) held by the Atomic Energy Society of Japan, Nagoya. Retrieved April 15, 2014 from http://www.aesj.or.jp/ ~ sed/forum/forum2004/sympo.htm.
Takagi, J., Schneider, M., Barnaby, F., Hokimoto, I., Hosokawa, K., Kamisawa, C., et al. (1997). Comprehensive social impact assessment of MOX use in light water reactors. Tokyo: Citizens' Nuclear Information Center. Retrieved April 15, 2014 from http://www.cnic.jp/english/publications/pdffiles/ima_fin_e.pdf.
Takahashi, T. (1993). Ikata Fukushima dai-ni genpatsu soshyou saikousai hanketsu (The supreme court's decision in the Ikata and Fukushima No. 2 atomic power plants trial). *Jurist*, 1017, 48–65. Retrieved April 15, 2014 from http://www.yuhikaku.co.jp/static_files/shinsai/jurist/J1017048.pdf.
Takahashi, S. (1998). *Sentan-gijyutsu no gyousei-houri (Administrative law thinking on advanced technology)*. Tokyo: Iwanami Shoten.
The Nagoya High Court Kanazawa Branch. (2003). Decision on Monju administrative lawsuit. http://www.page.sannet.ne.jp/stopthemonju/side/saiban/hanketuzenbun1.pdf.
Weinberg, A. M. (1972). Science and trans-science. *Minerva, 10*(2), 209–222. doi:10.1007/bf01682418.
Yamashita, R. (2003). Gyousei-hou-riron ni okeru genpatsu soshyou no igi (The meaning of atomic power plant trials in administrative law theory). *Jurist*, 1251. Retrieved April 15, 2014 from http://www.yuhikaku.co.jp/static_files/shinsai/jurist/J1251082.pdf.
Yasui, I. (2003). Monju no setchi kyoka mukou hanketsu (The judgment of nullity of Monju construction permit). Retrieved April 15, 2014 from http://www.yasuienv.net/MonjuMukou.htm.
Yoshioka, H. (1999, 2011). *Genshiryoku no shyakaishi (Social history of atomic energy)*. Tokyo: Asahi Shimbun.

Chapter 9
AIDS Patients Due to Transfusion of HIV Infected, Non-heat-treated Blood Products

Yoshiyuki Hirono

Abstract Many hemophiliacs have suffered from AIDS because particular lots of non-heat-treated blood products, used in the first half of the 1980s to treat hemophilia, were tainted with HIV. We needed to replace the non-heat-treated products with heat-treated products as soon as possible. Norway and Finland succeeded in this replacement, and sustained relatively little losses. Failure in this replacement in France, Canada and Japan led to a relatively great deal of harm. Japan delayed the implementation of relevant policy mainly because of the judgment of the AIDS research division of the Ministry of Health and Welfare that Japan should continue to use non-heat-treated blood products. One president and two ex-presidents (at that time) of a pharmaceutical company, which did not recall the risky non-heat-treated blood products, were sentenced to imprisonment without work. One bureaucrat was handed a suspended sentence of imprisonment without work. One doctor, who was a leading scholar in hemophilia and served as section chief of the AIDS research division, was judged not guilty because treatment using non-heat-treated blood products was a medical care standard of the time (MCST). There is an imbalance of responsibility since medical policies practically reflect the intention of the medical community. The theory of MCST lets a medical community remain behind advanced policy because it is immune from juridical responsibility. A juridical system based on the theory of MCST does not lead to rapid progress in medical policy for risk avoidance. We need to construct alternative theories to the theory of MCST, or other mechanisms than the juridical system.

Y. Hirono (✉)
The University of Tokyo, Tokyo, Japan
e-mail: yhirono@hps.c.u-tokyo.ac.jp

© Springer International Publishing Switzerland 2015
Y. Fujigaki (ed.), *Lessons From Fukushima*,
DOI 10.1007/978-3-319-15353-7_9

9.1 Introduction: Final Acquittal of Takeshi Abe

On April 25, 2005, Dr. Takeshi Abe died after what can be described as an eventful, controversial life. Abe won acclaim for his distinguished achievements in the study of hemophilia[1] and served as president of the Japanese Society of Hematology. Late in his life, however, he came under severe criticism for mishandling the response to the risk of HIV infection among hemophiliacs[2] and was charged with manslaughter through professional negligence in the death of a hemophiliac infected with HIV through contaminated blood products.

The circumstances of the charges against Abe in the death of the HIV-infected hemophiliac, or Patient J, are as follows. Patient J was injected with non-heat-treated blood products[3] three times—on May 12, June 6 and June 7, 1985—as a treatment for hemophilia by doctors at the Department of Internal Medicine I of Teikyo University Hospital. The non-heat-treated blood products were tainted with HIV.[4] As a result, Patient J later developed AIDS and was admitted to Teikyo University Hospital in October 1991. He died at the hospital in December of the same year.

In January 1996, the mother of the victim filed criminal accusations against Abe, who was the supervisor of the Department of Internal Medicine I of Teikyo University Hospital at the time. In September of the same year, the Tokyo District Public Prosecutors Office arrested Abe and brought charges of manslaughter through professional negligence against him. However, the Tokyo District Court (presiding judge: Toshio Nagai) acquitted Abe of the charges in March 2001.

This ruling provoked a storm of controversy (Sakurai et al. 2001). In April 2001, the Tokyo District Public Prosecutors Office appealed the case to the Tokyo High Court. At the first hearing of the appeal trial on November 29, 2002, the prosecution requested rejection of the district court's ruling and the conviction of Abe, arguing

---

[1] We bleed when we get cut, and if we are physically unimpaired, the bleeding stops after a while. This phenomenon, called hemostasis, is a process of blood coagulation by a complex chemical reaction. If it were a simple chemical reaction and proceeded easily, the blood would clot in the vessels of healthy people and could bring about an unfavorable result. This may be the reason why this process is complex, but at the same time, problems arise from such complexity. Many substances are involved in blood coagulation (blood coagulation factors I–XIII, consisting of proteins). If there is a problem with even only one of them, the entire process of blood coagulation is likely to break down. In fact, some people congenitally lack a blood coagulation factor. Once they start to bleed, their bleeding cannot be easily stopped, putting their life at risk. This is hemophilia. The cases of lacking Factor VIII are called hemophilia A and are the most common, accounting for 80 % of the total number of hemophiliacs (about 3,500–4,000 persons in Japan). The second largest group is hemophilia B patients who lack Factor IX (about 700 persons in Japan). Hemophilia is a sex-linked disorder, and most hemophiliacs are men. Women whose genes lack blood coagulation factors are called hemophilia carriers. Hemophiliacs are found at a rate of one per 10,000 persons (5,000 men) worldwide. In Japan, there are about 5,000 hemophiliacs.

[2] Acquired Immunodeficiency Syndrome: AIDS.

[3] To be precise, referred to as non-heat-treated and concentrated blood coagulation factor VIII products; product name: Cryoblin (Nippon Zoki Pharmaceuticals).

[4] Human Immunodeficiency Virus: HIV.

that the ruling failed to give consideration to the viewpoint of patients. Evidence examination was completed at the ninth hearing on December 16, 2003, and the trial was scheduled to be concluded on March 2, 2004. However, the defense petitioned for a stay of the trial proceedings on November 17, 2003.

That was because Abe was admitted to the hospital for a long stay twice between April and September 2003 because of heart and other health problems and was deemed to be incapable of communication after leaving the hospital (it is said that he was suffering from dementia). Keio University Hospital, which was commissioned to conduct a psychiatric test on Abe, provided the Tokyo High Court with a written expert opinion to the effect that Abe was in a state of non compos mentis and was insane. After an interview with Abe on February 16, 2004, the Tokyo High Court determined on February 23 that Abe was mentally unfit to stand trial and approved a stay of court proceedings. Following Abe's death on April 25, 2005, the Tokyo High Court (presiding judge: Yoshimasa Kawabe) invalidated the charges against him on May 13, 2005. This put an end to the trial without a final conclusion on the issue of criminal responsibility in the case.

The purpose of this chapter is to review Japan's response to the risk of HIV infection through tainted blood products from the perspective of the government's science policy and interpret the meaning of Abe's acquittal from the viewpoint of the study of science, technology and society.[5] First, I will provide an overview of the history of AIDS and HIV infection through tainted blood products, and then I will review the litigation cases related to the spread of HIV infection among hemophiliacs. After that, I will describe the history of the study of AIDS and the history of the response to the AIDS epidemic. Through international comparison, I will identify the notable features of Japan's response to the risk of HIV infection through tainted blood products. Through these processes, I aim to clarify the significance of Abe's acquittal from the perspective of the study of science, technology and society.

9.2 HIV Infection Through Tainted Blood Products

When we look at the history of medicine, we are reminded that the history of the world is one of a fight against infectious diseases. New infectious diseases emerge one after another. AIDS is one of such "emerging infectious diseases."

AIDS is presumed to have originated in Africa. HIV, which causes AIDS, is generally believed to be derived from the simian immunodeficiency virus (SIV), which infects primates. It is presumed that the initial cases of HIV infection were associated with the practice of eating chimpanzee meat.

---

[5] When considered as a social issue, the range covered by the issue of HIV infection through medical practices is broad. In addition to the case caused by the use of tainted blood products, which is discussed in this paper, HIV infection could also occur due to transfusion of tainted blood. Blood transfusion caused more serious cases in other countries, such as Belgium, the United States, Romania, France, Portugal and Switzerland. Regrettably, I omitted this issue from this chapter.

According to one theory, AIDS was brought from Africa to the United Kingdom by ship crews around 1958–59. Meanwhile, AIDS is presumed to have crossed the Atlantic with a Haitian volunteer soldier who fought in Africa when he returned home and to have arrived in the United States around 1974–1976 through travellers to Haiti.

A report on five cases of *Pneumocystis carinii* pneumonia made by the Centers for Disease Control and Prevention (CDC) in the June 5, 1981, issue of its *Morbidity and Mortality Weekly Report* (MMWR) represents the first official report on the discovery of the disease that later came to be known as AIDS (CDC 1981a, b). It was in the September 24, 1982, issue of the MMWR that the name AIDS was officially used for the first time (CDC 1982b).

In July 1982, the CDC reported a case of a hemophiliac with AIDS symptoms (CDC 1982a). This report suggested the possibility that donated blood and blood products were contaminated with HIV and hemophiliacs were exposed to the risk of infection.

The United States was the first advanced country to experience widespread HIV infection. In the United States, HIV infection initially spread among homosexuals. Subsequently, infection spread among drug addicts through the sharing of hypodermic syringes and among hemophiliacs and other people through the transfusion of HIV-tainted blood donated or sold by infected homosexuals and drug addicts and through the use of blood products created from contaminated blood. Then, infection started to spread through heterosexual contact. In Japan, which depended heavily on the United States for its supply of blood products, HIV infection first spread among hemophiliacs.

9.3 Litigation Cases Related to HIV Infection Through Tainted Blood Products

9.3.1 Case of HIV Infection—Patient K

One of the objectives of a court trial is resolving a specific dispute. It is impossible, in the first place, to expect a court trial to serve as an occasion to pass judgment on the policy response to the HIV infection through tainted blood products. In the HIV-tainted blood affair, criminal trials were held in connection with the death of two infected persons.[6] One is Patient J, whose case was mentioned in Sect. 9.1. The other is Patient K.

---

[6] In relation to the incident of HIV infection through tainted blood products discussed herein, civil actions were filed in Osaka in 1989 and in Tokyo in October the same year, to seek damages against the then Ministry of Health and the company that manufactured the tainted blood products. These civil cases were settled after then Minister of Health and Welfare Naoto Kan made an apology to the sufferers in February 1996, and the parities reached a settlement in March the same year.

After being diagnosed with esophageal varices, Patient K was treated with sclerotherapy. In the treatment which was conducted April 1–3, 1986, non-heat-treated blood products[7] were used to halt bleeding. As the blood products were contaminated with HIV, Patient K developed AIDS by around September 1993 and died in December 1995.

9.3.2 Filing of Charges Against Abe

As was mentioned earlier, in the case of Patient J, the Tokyo District Public Prosecutors Office brought charges of manslaughter through professional negligence against Abe in connection with his alleged inappropriate medical practice in September 1996. In addition, in October of the same year, the prosecution also brought charges of manslaughter through professional negligence against Akihito Matsumura, who was director of the biologics division of the Ministry of Health and Welfare at the time, in the deaths of Patients J and K for his alleged failure to promote a ban on and recall of non-heat-treated blood products.

In the case of Patient K, in March 1996, his wife filed a criminal accusation of murder against Renzo Matsushita, a former president of Green Cross Corporation, a pharmaceutical company which sold the HIV-tainted non-heat-treated blood products. Following the accusation, the Osaka District Public Prosecutors Office arrested and brought charges of manslaughter through professional negligence against Matsushita and two other former presidents of Green Cross, Tadakazu Suyama and Takehiko Kawano in October of the same year. Matsushita was president of Green Cross in 1983–1988, while Suyama served in the post in 1988–1993. Kawano was president of the company in 1993–1996. In April 1986, when the incident in question occurred, Matsushita was president, Suyama was vice president and Kawano was senior managing director.

In short, Abe and Matsumura faced criminal charges in the death of Patient J, while Matsumura, Matsushita, Suyama and Kawano did in the case of Patient K.

9.3.3 Chronology and Outcome of the Trials

I already mentioned the chronology and outcome of the trial of Abe in the death of Patient J (see Sect. 9.1). As for the trial of Matsumura, on September 28, 2001, the Tokyo District Court (presiding judge: Toshio Nagai) acquitted him of some of

---

[7] To be precise, referred to as non-heat-treated and concentrated blood coagulation factor IX products; product name: Christmassin (Green Cross Corporation). Green Cross Corporation was later merged into Yoshitomi Pharmaceuticals Industries. Movement toward reorganization has been continuing in the pharmaceutical industry. Green Cross Corporation's business is now undertaken by Mitsubishi Pharma Corporation.

the charges against him and sentenced him to one year in prison without work with a two-year suspension, compared with the three years' imprisonment without work demanded by the prosecution (December 28, 2000). The charges rejected were professional negligence charges in the death of Patient J.

In September 2001, Matsumura appealed the case to the Tokyo High Court, followed by an appeal by the prosecution in October of the same year. On March 25, 2005, the Tokyo High Court (presiding judge: Yoshimasa Kawabe) upheld the district court's ruling—the sentence of one year imprisonment without work with a two-year suspension on the charges relating to the death of Patient K and the acquittal in the death of Patient J—and rejected the appeals of both the defense and prosecution. On April 6, 2005, the defense appealed to the Supreme Court regarding the guilty decision relating to the death of Patient K, while on the same day, the prosecution abandoned its appeal. Under the Code of Criminal Procedure, an appeal to the Supreme Court is permissible only in a limited range of grounds such as a constitutional violation or a grave factual error that could affect the judgment. Although the Tokyo High Public Prosecutors Office expressed "regrets" at Abe's acquittal in the death of Patient J, it concluded after "examining the ruling from various viewpoints" that it was "inevitable to abandon the appeal because there was no appropriate ground on which to make an appeal." The Second Petty Bench of the Supreme Court (presiding judge: Yuuki Furuta) rejected his appeal on March 3, 2008. Thus, his conviction was finalized.

Next, I will describe the court proceedings relating to the three former presidents of Green Cross who stood trial in the HIV-tainted blood affair. The prosecution demanded three years' imprisonment without work against Matsushita and two years and six months' imprisonment without work against each of Suyama and Kawano. On February 24, 2000, the Osaka District Court (presiding judge: Mikio Miyoshi) sentenced Matsushita to two years in prison without work, Suyama to one year and six months in prison without work and Kawano to one year and four months in prison without work. The court determined that "it was gross negligence to fail to suspend sales of and recall non-heat-treated blood products while putting commercial interests first and ignoring the risk of HIV infection, even when it became possible to supply safe, heat-treated blood products."

All three defendants appealed to the Osaka High Court. On February 24, 2002, the second criminal division of the Osaka High Court (presiding judge: Ken Toyota) handed down a sentence of one year and six months in prison without work against Matsushita and a sentence of one year and two months in prison without work against Suyama. The appeal against Kawano had been invalidated following his death from illness in May 2001.

Matsushita and Suyama immediately appealed to the Supreme Court, but the Third Petty Bench of the Supreme Court (presiding judge: Toyozou Ueda) rejected their appeal on June 27, 2005. Consequently, their convictions were finalized.

9.3.4 Prospects for Final Decision on Criminal Responsibility

Although the Supreme Court has yet to issue a ruling in the HIV-tainted blood affair, we can summarize the judicial proceedings so far: the proceedings resulted in guilty verdicts in the death of Patient K and acquittal in the death of Patient J. There are several factors that may have led to different decisions in the two cases, one of which is the difference in the timing of the deaths of the two patients.

The rulings issued in these cases appear to be based on the theory that the period between July 1985 and February 1986 is a watershed in this affair. According to the theory, nobody can be held responsible for cases of HIV infection through non-heat-treated blood products that occurred before that period. Therefore, the use of non-heat-treated blood products in the case of Patient J on three occasions in 1985— on May 12, June 6, and June 7—was not seen as a guilty act while their use on April 1–3, 1986 was recognized as such. Why is the period between July and February 1986 viewed as a watershed (detailed explanations are to be provided in Sect. 8.5.3)?

9.4 History of the Response to the HIV-Tainted Blood Affair

9.4.1 Time Lag of Two Years and Four Months

It was around 1993–1996 when problems with Japan's response to HIV infection through tainted blood products in 1982–1985 attracted intense media attention. At that time, a "time lag of two years and four months"[8] was often mentioned in relation to Japan's response. What does that time lag refer to?

In modern medicine, a procedure known as whole blood transfusion—a transfusion of fresh blood from other people—was the first method of treatment of hemophilia. However, whole blood transfusion was problematic because fresh blood can be preserved only for a short period of time. In addition, if repeated several times, the procedure increases red cells in the patient's blood to an excessive level.

Second, the method of freezing the plasma component of blood and using it as necessary was developed. But this method, if repeated frequently, causes side effects such as high blood pressure.

Third, a treatment using cryoprecipitate, or cryo, was introduced. Cryo is a blood product prepared from a deposit containing blood-clotting factors that are produced when the plasma component is cooled to its freezing point. "Cryo" is a Greek word that means "cold."

---

[8] For example, "Kakusareta eizu" (Hidden AIDS), a book edited by the City Department of Mainichi Newspaper (1992; Diamond), contains a chapter under the heading of, "the truth behind the blank of two years and four months—the cause of the increase in the number of victims".

Fourth, a concentrated blood product was developed, an achievement made possible by advances in blood separation technology. This is a concentrate of blood-clotting factors collected from blood taken from a large number of people, ranging from thousands to tens of thousands of people.

Finally, a heat-treated concentrated blood product was developed as an alternative to existing, non-heat-treated blood products in order to prevent infection.

As was mentioned earlier, it was in July 1982 that the risk of HIV infection among hemophiliacs was first reported (see Sect. 9.2). The response to the HIV-tainted blood affair that I am discussing refers to actions taken in the period of two–three years that followed this report. Around July 1982, the treatment of hemophilia was shifting from the use of cryo to the use of (non-heat-treated) concentrated blood products. Concentrated blood products were preferred over cryo because they produced much more effect per unit and were also easier to handle.

In the United States, where the AIDS epidemic spread first among advanced countries, the response was relatively quick. On March 21, 1983, around six months after the first report on the risk of HIV infection among hemophiliacs, the use of heat-treated blood products was approved in the United States. However, it was not until July 1985 that the use of such products was approved in Japan.

If we take the position that the damage could have been minimized if Japan also introduced heated blood products from the United States immediately, it may be said that Japan's response lagged behind the U.S. action by some two years and four months. According to testimony by Atsuaki Gunji, Matsumura's predecessor as director of the biologics division of the Ministry of Health and Welfare, the government considered the possibility of importing heat-treated blood products as an extra-legal emergency measure.

9.4.2 Time Lag of 10 Months

However, Dr. Kaizo Kanuma took issue with the argument mentioned in the previous section and contended that the time lag of Japan's response to the risk of HIV infection through contaminated blood products was around one year (10 months, to be precise). His argument is as follows:

> The time lag of Japan's approval of heat-treated blood products is not two years and four months as is usually mentioned. The counting of the time lag should start from September 1984, by which time HIV had been identified as the virus that causes AIDS, risks associated with non-heat-treated products had become clear, antibody testing had become available, and major countries around the world had approved heat-treated blood products following the confirmation of the safety of such products in medical circles around the world. There was a time lag of around one year, (from September 1984 to July 1985), and this should be judged to constitute negligence (Kanuma 1998, 169).

As is shown above, there are differences of opinion as to how long Japan lagged in responding to the risk of HIV infection through tainted blood products. Below, I will elaborate further on this point.

9.4.3 First Milestone: January 1983—Proposal from Desforges and Recommendation from the National Hemophilia Foundation

In my opinion, there are three milestones of the response to the risk of HIV infection through tainted blood products. In June 1981, the existence of a new disease that would later be named AIDS became widely known in medical circles. In July 1982, the CDC reported on a case of a hemophiliac with AIDS symptoms. On December 5, 1982, the CDC announced that although the cause of AIDS was not definitely known, it was deemed to be an infectious disease and that it was necessary at the moment to take preventive measures similar to those taken against hepatitis B, to which the new disease was similar in some aspects.

A month later, in January 1983, the New England Journal of Medicine carried two relevant articles (Lederman et al. 1983; Menitove et al. 1983). Lederman et al. (1983) reported on the results of a lymphocyte test (T4:T8 ratio) on hemophiliacs. The results showed that hemophiliacs treated with cryo were less prone to immunodeficiency than those treated with concentrated blood products. In a commentary on the two articles in the same issue of the journal, Jane F. Desforges, a hemophilia expert, proposed that the treatment of hemophilia shift back from the use of (non-heat-treated) concentrated blood products to the use of cryo (Desforges 1983). Desforges' argument was as follows: The current treatment method is quite successful, so it would not be easy for either doctors or patients to abandon it. However, despite insufficient evidence, now is the time to radically change the treatment method. The fact that hemophiliacs are at risk of being infected with HIV is becoming increasingly clear. If the use of cryo can reduce this risk, the current method that uses intravenous injection at home should be revised.

Also in January 1983, a conference was held in Atlanta to discuss the relationship between AIDS and blood, with the participation of representatives from the CDC, the FDA, blood banks and pharmaceutical makers. Based on the results of debates at this conference, the National Hemophilia Foundation issued a recommendation concerning blood products (hereinafter referred to as the "1983 NHF recommendation"). The 1983 NHF recommendation called for the use of cryo to be given precedence with regard to the treatment of (1) newborn babies and infants up to four years old, (2) patients who had never been treated with (non-heat-treated) blood products and (3) mild cases of hemophilia.

As was mentioned earlier, at that time, there were two options—using cryo or non-heat-treated blood products—for treating hemophilia, and a shift from cryo to non-heat-treated blood products was the trend.

Whereas a dose of cryo was usually prepared from blood taken from a few persons, production of (non-heat-treated) concentrated blood products usually involved blood taken from a large number of people. The risk of infection would be far higher when (non-heat-treated) blood products were used than when cryo was used.⁹ For example, a (non-heat-treated) concentrated blood product involving blood from 100 people would have a HIV contamination risk 30–50 times higher than a dose of cryo. The higher risk of (non-heat-treated) blood products was corroborated by Lederman's research. Indeed, no case of a hemophiliac who was treated with cryo but not with (non-heat-treated) concentrated blood products and who was infected with HIV has been reported in either the United States or Japan.

Although (non-heat-treated) concentrated blood products were highly effective in treating hemophiliacs, they were found to have a high risk of HIV infection. Therefore, the risk could outweigh the benefits if mild cases of hemophilia for which less effective treatment would be sufficient were treated with concentrated blood products. To put it another way, the use of (non-heat-treated) concentrated blood products should have been permitted only for the treatment of serious cases of hemophilia which were otherwise highly likely to result in death due to bleeding and for which the benefit of high effectiveness in halting bleeding is significant.

In light of these circumstances, Desforges' proposal and the 1983 NHF recommendation were made with the aim of reversing the shift from cryo to (non-heat-treated) concentrated blood products and minimizing the risk.

In hindsight, it may be said that Desforges' proposal and the 1983 NHF recommendation were the right approach. When it comes to policy decision-making, the responsibility of policymakers for failing to resolve a problem is far heavier when they have not adopted a solution on the table than when no solution is available in the first place. Therefore, people involved in the response to the risk of HIV infection after Desforges' proposal were held accountable. That is why I characterize January 1983, when Desforges' proposal and the 1983 NHF recommendation were presented, as the first milestone.

However, it should be noted that what was known at that time was limited. The major known facts were (1) that a new disease which eventually came to be known as AIDS was discovered (June 1981), (2) that the average life expectancy of people with AIDS symptoms was 1–2 years, (3) that the disease was spreading among hemophiliacs (July 1982) and (4) that the disease was infectious (July 1982). We are reminded how little was scientifically known about HIV/AIDS in those early days.

9.4.4 Second Milestone: June 1983—Recommendation from the Congress of the World Federation of Hemophilia

On June 29, 1983, the Congress of the World Federation of Hemophilia (WFH) was held in Stockholm, and the following recommendations were issued.

---

⁹ It has been revealed that in Japan, some pharmaceutical companies sold cryo products made of blood taken from a large number of people.

1. At the moment, there is not sufficient evidence to recommend any change in the treatment of hemophilia. Therefore, doctors involved in the treatment should continue the current treatment using whichever blood product may be available.
2. It is necessary to more accurately compare the risks and merits of various treatment methods. It is also necessary to conduct a time-sequential study at an early date.

These recommendations, which are presumed to have been drafted by **Dietrich,** were adopted at the Congress although they drew opposition from the Netherlands and Switzerland. This is the second milestone.

The following scientific facts were discovered during the five-month period spanning the presentation of Desforges' proposal and the 1983 NHF recommendation and the adoption of these recommendations at the Congress of the WFH in Stockholm.

In March 1983, it was concluded that the spread of AIDS among hemophiliacs was caused by tainted blood products (CDC 1983a). At around that time, the CDC announced the criteria for the diagnosis of HIV infection (CDC 1983b). On May 20, 1983, a group led by Luc Montagnier at the Institut Pasteur of France isolated the virus from samples taken from AIDS patients and named it LAV (Bareé-Sinoussi et al. 1983).[10]

As a result of the evolution of scientific knowledge concerning AIDS, the necessary minimum conditions were set for doctors on the frontline of medical practice to diagnose HIV/AIDS. In addition, the discovery of a virus that causes AIDS increased the probability that AIDS was infectious. In relation to hemophilia, it was seen highly likely that the use of (non-heat-treated) concentrated blood products involved risks.

In the United States around that time, Travenol (which is now called Baxter) obtained approval for the production of heat-treated blood products. However, this was primarily intended as a measure to deal with the risk of hepatitis infection. In any case, on May 24, 1983, the U.S. Public Health Service expressed expectations that heat-treated blood products would help to reduce the risk of HIV infection while acknowledging that the cause of AIDS was unknown. Consequently, the use of heat-treated blood products emerged as the third option for treating hemophilia.

What is notable about the second milestone is that despite the accumulation of scientific knowledge, it was recommended that the current treatment method (use of non-heat-treated blood products) be continued.

---

[10] Gallo, a US researcher, also isolated a virus from samples taken from AIDS patients and named it HTLV-I. However, it was later found that this virus was not a causative virus of AIDS but was a cause of tropical spastic paraparesis, which is an endemic disease in the Caribbean coastal area, and adult T-cell leukemia (ATL), which is an endemic disease in Japan. Following Montagnier's achievement, Gallo discovered the true causative virus of AIDS and named it HTLV-III (Gallo et al. 1984).

9.4.5 Third Milestone: October–November 1984

Around one year and three months later, on October 13, 1984, the NHF changed the guideline for treatment and recommended the use of heat-treated blood products. The NHF stated that a shift to heat-treated blood products should be seriously considered based on the understanding that whether or not the agent that causes AIDS can be inhibited has not been verified. (The recommendation shall be hereinafter referred to as the "1984 NHF recommendation.") This is the third milestone. In the same month, the CDC expressed its support for the 1984 NHF recommendation.

The following scientific facts were discovered during the period of around one year and three months between the second and third milestones.

In September 1984, Abe sent 48 samples of blood taken from hemophiliacs to Professor Robert Gallo in the United States and commissioned him to conduct an antibody test. The test results showed that samples taken from 23 hemophiliacs, including two who had died, were infected with HIV (infection rate of approximately 48 %). The results were not published for around eight months, a fact that would later raise suspicions of a cover-up.

Also in September 1984, the International Congress of Virology concluded that AIDS is caused by LAV/HTLV-III. At the same time, the international committee on taxonomy of viruses of the International Union of Microbiological Societies adopted HIV as the unified name of the virus that causes AIDS.

On May 4, 1984, methods of HIV antibody testing were developed independently by Montagnier and Gallo. The development of the test contributed to an advance in knowledge concerning AIDS by making it possible to identify people who were somehow connected with HIV although not showing AIDS symptoms. However, questions remained over the relationship between a positive HIV reading and the development of AIDS symptoms.

In February 1984, in the United States, heat-treated blood products introduced to reduce the risk of hepatitis infection were also approved for application to hemophilia. On October 26, 1984, the CDC reported the verification of inactivation of HIV as well as hepatitis virus in heat-treated blood.

What are the facts that are known as of now (2014) but remained unclear at that time? To find out, let's take a look at how things have unfolded since then. As of the spring of 1984, little was known about the HIV infection rate or the AIDS incidence rate (Evatt et al. 1984). In the December 12, 1984, issue of MMWR, it was reported that most adults infected with HIV remained without AIDS symptoms for several years. At the first International AIDS Conference that was held in Atlanta on April 11, 1985, it was reported that 7–14 % of HIV positive people developed AIDS symptoms while 60–70 % remained without AIDS symptoms during the observation period (2–5 years). According to a survey report concerning blood provided by homosexuals and preserved, 5–19 % developed AIDS symptoms within 2–5 years after a positive HIV reading while long-term prospects for the development of the disease remained unclear. The January 17, 1986, issue of

MMWR carried a follow-up survey that covered a seven-year period from HIV infection, which indicated that the incubation period could be unusually long.

Let me sum up what was known and what remained unclear as of September–October 1984. First, the virus that causes AIDS was identified (Coffin et al. 1986). It became possible to identify people who were somehow connected with HIV (HIV positive) although not showing AIDS symptoms. It was found that 5–20 % of HIV positive people developed AIDS symptoms. Furthermore, heat-treated blood products started to attract attention. It was also starting to be recognized that the incubation period of AIDS was long.

However, a full statistical picture of the incubation period was not available at that time. Given that AIDS is presumed to have arrived in the United States in 1974–76 and that the average incubation period is 10 years, it would have been difficult until 1984–86 to fully identify the incubation period. A statistical picture of the incubation period was just starting to emerge in those days.

In relation to hemophilia, it was fairly possible to presume that hemophiliacs who were frequently treated with (non-heat-treated) concentrated blood products had a high probability of being HIV positive (the actual HIV-positive rate was approximately 50 %), that at least around 5–20 % of HIV-positive patients had developed or would develop AIDS symptoms in the future and that patients with AIDS symptoms would die within 1–2 years. In Japan, it may be said, it would have been easy around that time to presume that if hemophiliacs continued to be treated with non-heat-treated blood products, around 2,500, or half of all hemophiliacs in the country, would become HIV-positive and that 125–500 of them would develop AIDS symptoms and die.

Therefore, we may say that policymakers who were promoting the shift from non-heat-treated blood products to cryo by January 1983 were very far-sighted. Those who were not doing so at that point of time should also have done their utmost to switch to cryo or heat-treated blood products by October–November 1984.

9.5 International Comparison

9.5.1 Significance of a Time Lag of Around One Year

Kanuma argued that "there was a time lag of around one year, (from September 1984 to July 1985) and that "this should be judged to constitute negligence (Kanuma 1998, 169)." It would be difficult to understand the significance of the "time lag of around one year" by only looking at the situation in Japan. In this section, I will examine the significance based on international comparison.

Franceschi et al. (1995) published a comparative study of the incidence of AIDS associated with transfusion of blood and blood products. Meanwhile, Trebilcock et al. (1996) studied the relationship between the blood supply systems and HIV

Table 9.1 International comparison regarding the time lag of response to HIV infection through tainted blood products (1)

When the use of heat-treated concentrated blood products was made available		Time lag (month)
Germany	September 1984 (or 1978 for hepatitis)	−1
United States	October 1984 (or March 1983 for hepatitis)	0
Australia	November 1984	1
France	December 1984	2
United Kingdom	April 1985	6
Canada	May 1985	7
Japan	July 1985	9
Switzerland	May 1986 (or March 1983 for a limited volume of import)	19

Note This table shows the number of months from when the necessity to shift to the use of heat-treated blood products was made clear until when heat-treated blood products were made available. It indicates Japan's slowness in taking response. For the detailed explanation, refer to the text of this paper

infection through tainted blood in seven advanced countries. Based on the findings of Trebilcock et al. (1996), Glied conducted a further comparative study (Glied 1999). I will make my points while making references to these preceding studies.

Table 9.1 shows the dates when the use of heat-treated concentrated blood products, regardless of whether they were domestically produced or imported from abroad, were approved in various countries. To enable comparison with the "time lag of around one year" mentioned by Kanuma, the number of months from October 1984 (the timing of the third milestone) is also indicated in the table. Switzerland lagged behind in developing heat-treated blood products but started importing from Germany relatively early. From the table, the slowness of Japan's response is conspicuous.

Approval of heat-treated blood products alone would be insufficient to ensure safety. Full safety could be ensured only after the imposition of a ban on non-heat-treated blood products and the completion of the recall of such products in circulation. In this respect, it cannot be denied that the United States was relatively slow in taking action: the U.S. government banned production of non-heat-treated blood products in June 1985. As is widely known, Japan also failed to take action quickly in this respect. On August 28, 1986, the Japan Blood Products Association reported that the recall of non-heated-blood products in Japan had been completed. However, some pharmaceutical makers continued to sell non-heat-treated blood products until 1987, insisting that they were safe. Matsumura and three former presidents of Green Cross were found guilty in connection with the failure to halt sales of non-heat-treated blood products.

Table 9.2 shows the dates of the mandating of heat treatment of concentrated blood products as defined by Trebilcock et al. (1996) and Glied (1999), which is equivalent to a near total replacement of non-heat-treated blood products with heat-treated ones. This table indicates the number of months not only from October 1984

Table 9.2 International comparison regarding the time lag of response to HIV infection through tainted blood products (2)

When heat treatment of concentrated blood products was made mandatory		Time lag (month)	
United States	October 1984	0	0
Australia	January 1985	3	2
United Kingdom	June 1985	8	2
Canada	July 1985	9	2
Italy	July 1985	9	–
Denmark	October 1985	12	
France	October 1985	12	10
Japan	October 1985–February 1986	12–16	3–7
Germany	Heat treatment was not made mandatory but refund for non-heat-treated blood products was stopped		

Note This table shows the number of months until when non-heated treated products were completely replaced with heat-treated ones, as counted from when the necessity to shift to the use of heat-treated blood products was made clear, or from when heat-treated blood products were made available. It indicates Japan's slowness in responding. For the detailed explanation, refer to the text of this paper

but also from the dates when heat-treated blood products became available. It should be noted that the time lags as defined by Kanuma are not necessarily identical to the lags indicated in Table 9.2 because whereas Kanuma adopted the date of a blanket approval of heat-treated blood products as the end point of the time lag, the date of the mandating of heat treatment is used as the end point in Table 9.2. From Table 9.2, it cannot be denied that Japan's response was slow.

Next, I will look at the time lags from another viewpoint. In Table 9.1, October 1984 is the starting point of the time lag. However, it is not that countries started to take action overnight at some point of time. It would be natural to assume that countries gradually became aware of the infection risk related to non-heat-treated blood products by October–November 1984. Table 9.3 shows the time lags in terms of the number of months from the dates when risks involved in the blood supply system were recognized.

From Table 9.3, we can see that the United States, France and Canada were quick to respond to the CDC's report on cases of hemophiliacs with AIDS symptoms. In Japan, Gunji, who was director of the biologics division of the Ministry of Health and Welfare at that time, felt a sense of crisis after obtaining information on Desforges' proposal. In June 1983, the Ministry of Health and Welfare established a study group charged with assessing the situation of the AIDS epidemic (which was headed by Abe, who was a professor at Teikyo University at the time). The ministry consulted with the study group about the idea of imposing a blanket ban on imports of blood from the United States (the study group rejected this idea). Table 9.2 shows that although Japan was not as quick to recognize the risk as the United States, France and Canada, it was not such a laggard compared with Germany and Australia.

Table 9.3 International comparison regarding the time lag of the response to HIV infection through tainted blood products (3)

	When the risk was recognized	When heat treatment of concentrated blood products was made mandatory	Time lag (month)
United States	July 1982	October 1984	27
France	July 1982	October 1985	39
Canada	September 1982	July 1985	34
Germany	April 1983	Heat treatment was not made mandatory but refund for non-heat-treated blood products was stopped	
Australia	June 1983	January 1985	19
United Kingdom	September 1983	June 1985	21
Japan	January–June 1983	October 1985–February 1986	28–37

Note This table shows the number of months from when the risk of non-heat-treated blood products was recognized until when non-heated treated products were completely replaced with heat-treated ones. It indicates Japan's slowness in responding. For the detailed explanation, refer to the text of this paper

However, the time lag of Japan's response in terms of the time passed between the recognition of the risk and the action to deal with it was as long as 37 months based on the assumption that the risk was recognized in January 1983 and the action was completed in February 1986 (a maximum time lag case). Meanwhile, based on the assumption that the risk was recognized in June 1983 and the action was completed in October 1985 (a minimum time lag case), the time lag was 28 months.

Given that the United States was the first country to experience the AIDS epidemic and so faced difficulty figuring out how to deal with it, Japan's response time, which was as long as the U.S. response time in a minimum time lag case, can in no way be characterized as short. Australia and the United Kingdom took action 19–21 months after recognizing the risk, while learning from the U.S. response. In a minimum lag time case, Japan's response was quicker than the action of France and Canada. In a maximum time lag case, however, it took almost the same period of time for Japan to take action as France and Canada. Even if it is accepted that Japan's response was as quick as the French and Canadian response, it does not constitutes grounds for exempting Japanese policymakers at that time from the liability for failing to take action quickly. In France, Dr. Garretta, who was responsible for overseeing administrative affairs related to blood supply, was sentenced to four years in prison. A response as good as the French and Canadian action means a poor achievement.

9.5.2 Overall Situation

Now, I will conduct an international comparison in terms of the HIV infection rate among hemophiliacs. Overall, the infection rate is high for Romania, the United

States, France, Spain and Portugal, while the number is low for the Netherlands, Sweden, Germany, Italy and Belgium. With regard to infection through blood products alone, the infection rate is high for Spain, the United States and Austria and low for Belgium, Switzerland, Italy, the Netherlands and Sweden (Franceschi et al. 1995).

Table 9.4 shows the scale of the AIDS epidemic in individual countries based on data published by the WHO in 1994 and other information. Of course, the infection rate from treatment with non-heat-treated concentrated blood products (the ratio of HIV-positive hemophiliacs to all hemophiliacs) has increased by now to 60 % in the United States and to 45–50 % in Japan, for example. However, I use the WHO data published in 1994 because updated data concerning some regions are not available and the most recent data comparable across a large number of countries are those for 1994. Another reason is that I assume there has been no change in the general trend since then.

It should be noted that Table 9.4 incorporates data from "Per-Capita Gross Domestic Product Data" in 1994 by Maddison (1995).[11] Where data for 1994 are not available, I use data for 1992 (marked with *). For convenience's sake, countries are categorized by per capita-GDP into Group A—countries with per-capita GDP of 10,000 dollars (Geary-Khamis dollars) or higher—and Group B—countries with per-capita GDP of less than 10,000 dollars. Countries in each group are arranged in top-down descending order by the infection rate.

As countries in Group B are poor, concentrated blood products were not used widely there in the first place. As a result, the infection rate from treatment with non-heat-treated concentrated blood products is generally low in those countries.

In Group A, the infection rate is high for the United States, Spain, Canada, Japan and Ireland and low for South Korea, Finland, Norway, Belgium and New Zealand. Norway, Poland and Finland are well known as countries that avoided the tragedy of HIV infection through tainted blood products because of their proactive efforts to shift to cryo. The Netherlands, which strongly opposed the issuance of the WFH's recommendations at its Congress in Stockholm, also proactively switched to cryo, resulting in a relatively low infection rate in the country.

Scotland banned imports of non-heat-treated products from the United States from 1982 onward and shifted almost fully to domestically produced non-heat-treated blood products. As a result, the infection rate in Scotland at the end of 1984 was 15.6 % (12 infected patients out of the total of 77 patients), and all HIV positive patients there are presumed to have been infected before 1982 (Kanuma 1998, 80). In South Korea, domestic supply accounted for almost all blood products used there because of a traditional belief related to blood. As was shown above, the infection rate is generally low in countries which had never imported blood products from high-risk countries and those which immediately banned imports

---

[11] Maddison (1995). Monitoring the world economy, 1820–1992. Translated as Sekai keizai no seichōshi 1820–1992 (2000). Japanese translation supervised by Kanamori, H. Tokyo: Toyo Keizai.

Table 9.4 HIV infection through tainted blood products, by country

Country name	Number of HIV positives	Infection rate (%)	GDP per capita
A			
United States	9,000	50.10	22,569
Spain	1,147	43.60	12,544
Canada	653	37.90	18,350
Japan	1,792	37.50	19,505
Ireland	101	33.80	12,624
Germany	1,377	33.40	19,097
France	1,300	32.50	17,968
Portugal	129	27.60	11,083
United Kingdom	1,227	24.50	16,371
Italy	802	23.30	16,404
Greece	193	20.90	10,165
Austria	93	18.80	17,285
Denmark	90	18.00	19,305
Switzerland	67	16.20	20,830
Australia	260	16.10	17,107
Netherlands	170	14.00	17,152
Sweden	100	12.50	16,710
New Zealand	28	10.00	15,085
Belgium	40	7.70	17,225
Norway	21	6.30	18,372
Finland	2	0.90	14,779
Republic of Korea	2	0.60	10,010*
B			
Brazil	1,255	32.90	4,862
Argentina	198	21.00	8,373
India	29	14.50	1,348*
Venezuela	83	11.20	8,389
South Africa	89	11.10	3,451*
Peru	10	10.75	3,232
Thailand	11	5.60	4,694*
Romania	18	4.50	2,565*
Hungary	28	2.80	5,638*
Bulgaria	7	1.70	4,054*
Poland	16	0.90	4,726*
Bangladesh	0	0	720*
Philippines	0	0	2,231*

Dates marked with * are as of 1992 and others are as of 1994 from high-risk countries. Even so, although France banned imports from the United States at a relatively early date, it was unable to keep the infection rate low because it was already becoming a high-risk country itself.

In hindsight, countries that succeeded in containing the risk of HIV infection through tainted blood products were those that proactively shifted to cryo in response to Desforges' proposal and the 1983 NHF recommendation. The successful group of countries also includes those which promptly banned blood product imports from high-risk countries and those which were self-sufficient in supply of blood products. Countries that switched to heat-treated blood products relatively early also successfully contained the risk of HIV infection. How should Japan's response be evaluated?

9.5.3 Response by Japan and Other Countries

How did Japan respond to Desforges' proposal and the 1983 NHF recommendation? Japan did little more than establish a study group charged with assessing the situation of the AIDS epidemic at the Ministry of Health and Welfare by June 1983, which reflected the sense of crisis felt by Gunji, who was director of the biologics division of the Ministry of Health and Welfare at that time, after obtaining information on Desforges' proposal. Meanwhile, in February 1983, the Ministry of Health and Welfare approved the use of hypodermic syringes at home by hemophiliacs, a measure which in effect promoted the use of non-heat-treated blood products. The Ministry of Health and Welfare's idea of suspending imports from the United States was rejected by the study group. In short, a shift to cryo was not even discussed as a realistic policy. A conclusion as to whether to shift to cryo was reached at the first meeting of the study group's subcommittee on blood products[12] in September 1983. The subcommittee rejected the plan to shift to cryo, arguing that "there are limits to the expansion of the scope of applications of cryo, so it is difficult to abandon the use of concentrated blood products as the mainstay treatment." As a result, Japan did not follow the recommendation for using cryo for patients for whom its treatment effect would be sufficient in order to reduce the risk of HIV infection.

What was Japan's response to the decision made by the Congress of the WFH in Stockholm in June 1983? Until that time, Abe was calling attention to the infection risk and trying to promote efforts to contain it. However, following the WFH's decision that the existing treatment should continue, it is said that his enthusiasm to

---

[12] The subcommittee on blood products was composed of the following members (titles omitted): Mutsumi Kazama (Teikyo University: Chairperson); Tadashi Kamiya (Nagoya University); Hidehiko Saito (Saga Medical School); Eiichi Tokunaga (Japan Red Cross Society); Hiroshi Nagao (Kanagawa Children's Medical Center); Hiromu Fukui (Nara Medical University); Junichi Yasuda (National Institute of Health); Junichi Yada (Tokyo Medical and Dental University); Kaneo Yamada (St. Marianna University School of Medicine).

take action weakened (Shiokawa 2004, 227). The Netherlands, which had drastically reduced imports of blood-clotting factors early in 1983, strongly opposed the WFH's decision. In the United Kingdom, France and Japan, the use of non-heat-treated blood products increased following the decision.

In a sense, Japan's response to international recommendations and decisions was inconsistent. While Japan did not follow the 1983 NHF recommendation, it complied with the WFH's decision in Stockholm. To put it another way, Japan had an inclination to stick with non-heat-treated concentrated blood products, rather than switch to cryo. In that sense, Japan's response was consistent (ironically consistent in increasing the risk).

As Japan consistently chose the wrong paths, the infection rate rose, as was the case with France and Canada, both of which also made many missteps.

9.6 Significance of the Acquittal of Abe as Viewed from the Perspective of Science, Technology and Social Study

The legal doctrine that the court relied on when acquitting Abe is based on what is generally called the new theory of negligence.[13] "The accused in this case would be held criminally responsible in cases where the accused had chosen the medical practice that involved a greater risk as compared to the benefits derived therefrom, even though an ordinary hemophilia doctor who had been placed in the position of the accused at that time would have never made such choice" (from the text of the judgment of the Tokyo District Court).

We have already confirmed earlier in this paper that an ordinary hemophilia doctor would have considered shifting from the use of non-heat-treated blood products to the use of cryo or heat-treated blood products in October–November 1984. "Even at the time of the incident, it had been possible to foresee that the use

---

[13] With regard to negligence, several legal theories have been advocated, namely, the former theory of negligence, the new theory of negligence, the new version of the new theory of negligence, and the new version of the former theory of negligence. According to the former theory of negligence, in cases where a person failed to foresee the occurrence of a crime although he/she could have foreseen such result if he/she had kept his/her mind attentive, the person should be punished for the crime (the foreseeability of the result is the essential requirement for finding negligence). The new theory of negligence requires failure to avoid the result despite the possibility to avoid it, such as a breach of the general rules in civil life or a breach of the duty of care, in addition to the foreseeability of the result. A crime is less likely to be established under the new theory of negligence because the scope of acts that would constitute a crime is narrowed due to the additional requirement. Another theory attempts to relax the requirement of foreseeability, arguing that a vague sense of unease, instead of a certain foreseeing, would be enough to establish a crime. This is called the new version of the new theory of negligence (the theory of uneasiness). Meanwhile, the new version of the former theory of negligence requires foreseeability as an objective condition, rather than a person's subjective or psychological state.

of non-heat-treated blood products imported from abroad could cause hemophiliacs to be infected with HIV and then to develop AIDS and finally die, and it is found that the accused himself was actually aware of such risk. In other words, the accused did not assess that there was no risk of causing the result but he assessed that there was a risk of causing the result but the risk was low" (from the text of the judgment of the Tokyo District Court).

Let us suppose that there was a hemophiliac who was bleeding seriously at the time of the incident. However, at that time, the volume of production of cryo was inadequate. Furthermore, heat-treated products had not been approved and therefore had not been put on sale until July 1985 for Factor VIII and December 1985 for Factor IX, respectively. In addition, it was during the period from October 1985 to February 1986 that heat treatment was made mandatory. Given these facts, it was difficult to say that the conditions for the shift to cryo had been satisfied by July 1985.

Cryo was not easily available, and heat-treated blood products would not be put on sale until after a few months. Under such circumstances, it would have been unrealistic for an "ordinary hemophilia doctor" to wait for safe, heat-treated blood products to be released and leave the patient untreated until then. Assuming so, comparison should be made between the risk arising from the use of non-heat-treated blood products and the benefit derived from the treatment with the use of these products. As reviewed earlier, there was a risk but it was (thought to be) not so high (at the time of the incident). Stopping the bleeding by using non-heat-treated blood products was indeed beneficial. In that case, if an ordinary hemophilia doctor attached importance to the benefit and chose to use non-heat-treated blood products, this choice cannot always be judged to be an unreasonable decision.

"We must say that there is still a reasonable doubt to be cleared if we were to find that an ordinary hemophilia doctor, who had been placed in the position of the accused at the time of the incident, would have refrained from using non-heat-treated blood products upon facing the bleeding that is disputed in this case. Therefore, the accused cannot be found to be criminally responsible for the crime of causing death or injury through negligence in the pursuit of social activities, for which he is charged in this case." "Consequently, we render a verdict of not guilty against the accused under Article 336 of the Code of Criminal Procedure and make a judgment as indicated in the main text" (from the text of the judgment of the Tokyo District Court).

Given the circumstances where it was difficult to acquire cryo and heat-treated blood products were not yet on sale, according to the new theory of negligence, it would be natural for Abe to be exempted from criminal responsibility for "having given contaminated, non-heat-treated blood products to Patient J on three occasions in 1985 (May 22, June 6 and June 7)," whereas Abe would be found guilty for "having given non-heat-treated blood products to Patient K on April 1–3, 1986."

The new theory of negligence might lead to such conclusion. The defense counsel argued as follows. "When it comes to a medical practice performed by a doctor, the medical care standard of the time would be tantamount to a law that the doctor should comply with in providing medical care. Even if the medical care standard is wrong when viewed in retrospect, a medical practice performed in compliance with this standard is lawful. According to the medical care standard for treatment of hemophilia at the time of the incident, the use of non-heat-treated blood products must be continued. Therefore, it is obvious that the accused, who was a hemophilia doctor, cannot be held responsible for negligence for having performed the medical practice in compliance with the medical care standard of the time." In response, the Tokyo District Court held that the defense counsel's argument cannot be accepted "if they mean to say that the fact that the majority of ordinary hemophilia doctors at the time of the incident in Japan continued to use non-heat-treated blood products directly proves that the act for which the accused has been charged in this case was in compliance with the medical care standard of that time, and ultimately leads to the finding that there was no negligence on the part of the accused." Nevertheless, the court's ruling seems to be basically in line with the defense counsel's argument.

As reviewed in Sect. 9.5 above, Japan's response was slow by international standards in this area. However, even if the medical care standard that prevailed at some time in the past is wrong when viewed in retrospect, a medical practice performed in compliance with this standard is supposed to be lawful. It is indeed questionable to condemn an act that was committed in the past by applying the present day's knowledge retrospectively. Many people argue this point. "We must not consider the past decision made by the parties in charge based on how we look at things and what we know now (Kanuma 1998, 149)." "It is totally unreasonable if we retroactively judge the persons who were in charge of taking measures at the time of the incident in the light of the advanced knowledge available today (Shiokawa 2004, 227)." I do agree with these arguments.

However, it is not always irrational to try to attribute responsibility to the persons in charge in Japan at that time by comparing their decision with the decisions made by the persons in charge in other countries of that time. As already reviewed above, Japan's response was generally slow. Japan should have taken measures such as shifting to the use of cryo without delay, avoiding the use of blood products imported from high-risk countries if possible, and making the use of heat-treated blood products mandatory at an early date, while watching the movements in other countries, and yet, none of these measures was achieved. Even granting that those persons in charge in Japan at that time are not legally responsible, I would say that they do have the responsibility in the context of science policy.

In the case where the Japanese standard fell short of international standards, if the new theory of negligence were to be adopted and the persons in charge at that time were to be exempted from responsibility on the grounds of such shortage in the Japanese standard, this could mean that the slower the advancement of the Japanese medical care standard for treating a disease is, the more desirable it is for individual persons who wish to be exempted from legal responsibility. In other words, the approach based on the new theory of negligence could unavoidably result in approving or increasing the lag in science policy.

From the perspective of science, technology and social study, it cannot be denied that Japan was slow in taking measures against HIV infection through tainted blood products. If so, what should be done now is to improve the system so that an appropriate science policy can be carried out as soon as possible in the event that the same kind of incident unfortunately takes place in the future. However, for the reasons explained herein, it seems impossible to expect the legal doctrine based on the new theory of negligence to serve to guide or correct a science policy. The future challenge for us, at the interface between science and law, is to create a legal doctrine that will go beyond the new theory of negligence, or if law is incapable of guiding a science policy, to find a way to promote an appropriate science policy.

References

Barreé-Sinoussi, F., Chermann, J. C., Rey, F., Nugeyre, M. T., Charmaret, S., Gruest, J., et al. (1983). Isolation of a T-lymphotropic retrovirus from a patient at risk for acquired immunodeficiency syndrome (AIDS). *Science, 220*, 868.
CDC. (1981a). Kaposi's sarcoma and *Pneumocystis* pneumonia among homosexual men—New York City and California. *Morbidity and Mortality Weekly Report, 30*, 305–308.
CDC. (1981b). Follow-up on kaposi's sarcoma and *Pneumocystis* pneumonia. *Morbidity and Mortality Weekly Report, 30*, 409–410.
CDC. (1982a). Epidemiologic notes and reports *Pneumocystis carinii* pneumonia among persons with hemophilia A. *Morbidity and Mortality Weekly Report, 31*(27), 365–367. Retrieved July 18, 1982, from http://www.cdc.gov/mmwr/preview/mmwrhtml/00001126.htm.
CDC. (1982b). Current trends acquired immunodeficiency syndrome (AIDS): Precautions for clinical and laboratory staffs. *Morbidity and Mortality Weekly Report, 31*(43), 577–580. Retrieved July 18, 1982, from http://www.cdc.gov/mmwr/preview/mmwrhtml/00001183.htm.
CDC. (1983a). *The case definitions of AIDS used by CDC for epidemiology surveillance*. Atlanta: CDC.
CDC. (1983b). Current trend prevention of acquired immunodeficiency syndrome (AIDS): Report of inter-agency recommendation. *Morbidity and Mortality Weekly Report, 32*, 101–103. Retrieved July 18, 1983, from http://www.cdc.gov/mmwr/preview/mmwrhtml/00001257.htm.
Coffin, F., Haase, A., Levy, J. A., Montagnier, L., Oroszlan, S., Teich, N., et al. (1986). Human immunodeficiency viruses. *Science, 232*, 687.
Desforges, J. F. (1983). AIDS and preventive treatment in Hemophilia. *New England Journal of Medicine, 308*, 94–95.
Evatt, B. L., Ramsey, R. B., Lawrence, D. N., Zyla, L. D., & Curran, J. W. (1984). The acquired immunodeficiency syndrome in patients with hemophilia. *Annals of Internal Medicine, 1000*, 499.
Franceschi, S., Dal Maso, L., & La Vecchia, C. (1995). Trends in incidence of AIDS associated with transfusion of blood and blood products in Europe and the United States, 1985–1993. *BMJ, 311*, 11534–11536.
Gallo, R. C., Salahuddin, S. Z. Popovic, M., Shearer, G. M., Kaplan, M., & Haynes, B. F. (1984). Frequent detection and isolation of cytopathic retroviruses (HTLV-III) from patients with AIDS and at risk for AIDS. *Science, 25*, 840.
Glied, S. (1999). The circulation of the blood: AIDS, blood and the economics of information. In: E. Feldman, & R. Bayer, (eds.), *Blood Feuds: AIDS, blood, and the politics of medical disaster* (pp. 323–348). Japanese edition: Glied, S. (2003). Ketsueki no junkan—Ketsueki to eizu to jōhō no keizaigaku (A. Yamashita, Trans.) In Feldman, E. & Bayer, R. (eds.). *Ketsueki kuraishisu: ketsueki kenkyū to HIV mondai no kokusai hikaku* (T. Yamada & S. Miyazawa, Eds., pp. 253–279). Tokyo: Gendaijinbun-sha.
Kanuma, K. (1998). *Yakugai eizu saikō (Review on the HIV-tainted blood scandal)*. Tokyo: Kadensha.

Lederman, M. M., Ratnoff, O. D., Scillian, J. J., Jones, P. K., & Schacter, B. (1983). Impaired cell-mediated immunity in patients with classic hemophilia. *New England Journal of Medicine, 308,* 79–83.

Maddison, A. (1995). *Monitoring the world economy,* 1820–1992. Paris: OECD Development Center. Japanese edition: OECD Development Center (2000). *Sekai keizai no seichōshi 1820–1992* (H. Kanamori, Trans.). Tokyo: Toyo Keizai.

Menitove, J. E., Aster, R. H., Casper, J. T., Lauer, S. J., Gottschall, J. L., & Williams, J. E. (1983). T-lymphocyte subpopulations in patients with classic hemophilia treated with cryoprecipitate and lyophilized concentrates. *New England Journal of Medicine, 308,* 83–86.

Sakurai, Y., Yamashina, T., Oi, A., Hama, R., Kai, K., & Shimizu, T. (2001). *Yakugai eizu "muzai hanketsu," doushitedesuka?* (HIV-tainted blood case ended up with an acquittal ruling. Why?). Tokyo: Chuokoron Shinsha.

Shiokawa, Y. (2004). *Watashi no "Nihon eizushi" (My version of "Japanese AIDS history").* Tokyo: Nippon-Hyoron-Sha.

Trebilcock, M., Howase, H., & Daniels, R. (1996). Do institutions matter? A comparative pathology of the HIV-infected blood tragedy. *Virginia Law Review, 82,* 1407–1492.

Chapter 10
Winny Criminal Case: How Have Controversial Science, Technology, and Society Problems Been *Solved* While Avoiding Conflicts?

Masahi Shirabe

Abstract This chapter analyses how the Winny criminal case developed and how people as well as the copyright protecting body defined the issues and situations surrounding it and *solved* the issues. Winny, pure peer-to-peer file sharing software as an application of advanced software technology components might lead to software innovations, while it has been a tool to facilitate illegal file sharing. Therefore, since the developer of Winny was arrested for being in charge of aiding and abetting copyright infringement, although the Supreme Court acquitted him seven years later, he had attracted praise from the software engineering community, law specialists, and citizens as well as censure in the process. In response to his arrest, the following interlinked problems had brought to citizens, courts and the copyright protecting body respectively: (1) building social consensus on acceptable software development, (2) establishing a landmark precedence on aiding and abetting copyright infringement, (3) revising the Copyrights Act to cope with online piracy. However, the link between the problems bridged by Winny were broken into pieces as the key decision makers closed themselves off from other problem areas and our society gradually shifted its attention away from the case. As a result, the problems as a whole went by the wayside although each problem obtained its local optimum. Therefore, our society did not have any clues about what socially acceptable software development is, but it returned to a stable condition.

10.1 Introduction

Dr. Isamu Kaneko was a talented Japanese computer software engineer who passed away on July 6, 2013. He received both praise and censure during his life.

M. Shirabe (✉)
Tokyo Institute of Technology, Tokyo, Japan
e-mail: shirabe.m.aa@m.titech.ac.jp

© Springer International Publishing Switzerland 2015
Y. Fujigaki (ed.), *Lessons From Fukushima*,
DOI 10.1007/978-3-319-15353-7_10

He first came to the attention of the software engineering community in 2000 when he joined a project funded by The MITOH Program[1] of the Information-technology Promotion Agency of the Ministry of Economy, Trade, and Industry (IPA/METI). At roughly the same time, he developed and released 3D physical simulators as freeware that were extolled in the software engineering related community. Professor Jun Murai, who is well known as the father of the Internet in Japan, lamented Dr. Kaneko's death as his talents were confirmed, as follows.[2]

> Dr. Isamu Kaneko was a valuable pioneer and hero in the field of software engineering.
>
> Although he encountered difficulties, these brought many supporters and friends to him. He was quoted as saying that he resumed activities to achieve his new dreams; thus, I expected extraordinary results from him.
>
> I take my hat off to people like Professor Hiraki, Professor Inaba, and Mr. Dan for the contributions to developing such favorable environments for Dr. Kaneko.
>
> Now, it is our heartfelt mission to understand, develop, and pass on his technology within the spirit in which it was intended.
>
> I would also like to make an ironclad promise that we intend to shed light on the social factors behind the difficulties he encountered to achieve a society where his spirit can be dynamically engendered.
>
> His guiding light continues to shines in our consciousness.
>
> We are thinking of him and praying for him at this time of loss.

Meanwhile, the *difficulties* described above are the results of his actions, which gained him recognition in our society. He developed and released file-sharing software called Winny, which was designed to make it difficult to detect who uploaded files on its peer to peer (P2P) network. Many people used the software to illegally share copyrighted content and software. He was consequently arrested for being in charge of aiding and abetting copyright infringement,[3] although the Supreme Court acquitted him seven years later.[4] He attracted praise from the software engineering community, law specialists, and citizens as well as censure in this process.

Winny as an application of advanced software technology components might lead to software innovations, while it has been a tool to facilitate illegal file sharing, which has recently started to calm down in Japan, as will be explained later.[5]

---

[1] "This program aims to discover and develop outstanding human resources called Super Creators. Specifically, these are persons possessing creative ideas and skills for achieving software innovation and who can put these ideas and skills to use." (Retrieved August 5, 2014, from http://www.ipa.go.jp/english/humandev/third.html).

[2] http://itpro.nikkeibp.co.jp/article/NEWS/20130707/489582/ (Retrieved August 5, 2014).

[3] The arrest of two users who uploaded copyrighted files apparently directly triggered his arrest.

[4] Supreme court decision (Retrieved August 5, 2014, from http://www.courts.go.jp/hanrei/pdf/20111221102925.pdf).

[5] According to the survey by the Contents Overseas Distribution Association (Retrieved August 5, 2014, from http://www.meti.go.jp/meti_lib/report/2010fy01/E001204.pdf), the percentage of people (over 15) using file-sharing software declined from 9.1 to 5.8 % in 2010. Moreover, middle and high school students who ceased to use such software cited the revision of the Copyright Act as its number two reason (21.9 %). In addition, it can be presumed that rapid diffusions of digital distribution and smartphones as well as price reduction of digital content has decreased the percentage even lower in the last few years.

When I think about Winny, I always remember a local news article[6] on a 2002 EASST meeting at York.

> **Think through the appliance of science**
>
> STEPHEN LEWIS finds out why we don't need to be scared of science - but we do need to be careful with it.
>
> JUST imagine it. Some crazy scientist comes up with a wacko idea for a souped-up new form of personal transport that can whisk you effortlessly from place to place in a tenth of the time normally required. The only drawbacks: it relies on the controlled explosion of a highly inflammable liquid for power; it has a side-effect of slowly poisoning the air we breathe, and it's so fast it is dangerous. Hundreds, no thousands, will be killed every year using it.
>
> It would never be allowed, would it? Of course it is! It's called the car.
> "If somebody tried to introduce a technology where you pump petrol today, it would never get passed!" says Steven Yearley with a dry, slightly donnish smile.
>
> (snip)

Although there has been deep-rooted criticism against prosecutors' extensions of the concept of aiding and abetting copyright infringement, were software engineers, in consideration of innovations in the future, to deem Dr. Kaneko's ideas *crazy*? Or, were people who accused Dr. Kaneko and his Winny merely scared of cutting-edge software engineering? Whatever the case, the Winny criminal case then gripped the nation's attention.

This chapter analyses how the Winny criminal case developed and how people as well as the copyright protecting body defined the issues and situations surrounding it. Then, ways of *solving* these issues by critically analyzing the copyright system in Japan will be discussed.

10.2 What Is Winny?

Winny is pure P2P file sharing software. File sharing software systems are comprised of pure P2P, hybrid P2P, and client-server systems.

Files and their search tags are typically stored on a server in client-server systems. Their users have to access servers to upload/download files. There are technological challenges for client-server systems to enhance computational and communications capacities much more for mass users as transactions in the servers are likely to be intensive.

P2P systems have the potential to solve such capacity problems. That is, as files are separately stored in nodes (i.e., computers) that run P2P file sharing software, computational/communication loads are balanced among nodes. Consequently, if

---

[6] Retrieved August 5, 2014, from http://www.yorkpress.co.uk/archive/2002/07/31/7922885. Think_through_the_appliance_of_science/.

super-large-scale file sharing services are socially required, the development of P2P file-sharing software can be justified.[7]

Although search tags are stored on servers in hybrid P2P systems, they are separately stored on nodes in pure P2P systems. Thus, access logs that might reveal the footprints of illegal file uploaders are not recorded centrally on pure P2P systems unlike those on hybrid P2P and client-server systems, and it is hard to block illegal file distributions by using search tags.

Winny is more than pure P2P file sharing software. It appears to be designed for illegal file sharing. Winny stores and transmits all files in small-encrypted cache files. There is no information about upload nodes like uploaders' IDs in cache files. As nodes relay files, it is difficult to identify upload nodes. Moreover, as cache files are stored in relaying nodes, it is impossible to distinguish the upload node from other relaying nodes after files are transmitted. Consequently, illegal file uploaders are strongly protected from being identified.

It is important to remember that although Winny has such *dark* specifications, it is also excellent software in terms of the quality of file sharing services. For example, it enhances the efficiency of file sharing to store cache files in relaying nodes. That is because the number of nodes storing a file becomes larger with the popularity of the file so that users can download the file from nearby relaying nodes. Interestingly, this specification is also useful for protecting the anonymity of illegal file uploaders. Although there are many other devices that offer better file sharing services than Winny, Winny is software that is evolving in a direction that conforms to the expectations of society.

Then, what are the main problems with Winny?

Winny was *intentionally* designed for illegal file sharing. That does not mean that its developer's primary purpose was to promote illegal file sharing. However, evidence for such an intention can be singled out due to the existence of technology components and absence of another component.

The first technology component is encryption. File encryption to anonymize users would not be necessary, unless software had purposes such as protecting free speech under brutal dictatorships, which Freenet (http://freenetproject.org/) was designed for. Winny does not have such a purpose[8] according to Dr. Kaneko's messages on a bulletin board system (BBS) for exchanging messages with its users. If so, encryption is only a factor that burdens its nodes. Thus, this wasteful implementation of encryption could be taken as strong but indirect evidence of his intentions to invite illegal file uploaders. However, it is noteworthy that encryption is not very important to protect illegal file uploaders from the danger of arrest in this regard. Unless police had then broadly monitored communication over the Winny

---

[7] For example, although Skype is not P2P file sharing software, it used P2P technology.

[8] Dr. Kaneko and his lawyer claimed that his release of Winny was a social verification experiment of a secured communication system like Freenet, and the court acknowledged it. But, as the encryption technology used in Winny was an established one, it was not very easy to justify its use in a *verification experiment*.

network, which has been illegal in Japan, they could hardly arrest illegal file uploaders under our Copyright Act, which was irrelevant to encryption.

Another technological component was devices to prevent file transactions (Ootani 2004a). Winny had no devices to block illegal file transactions. Or, there were no devices to suppress illegal acts by file uploaders. For example, if Winny had assigned a unique ID to each uploaded file,[9] users would have hesitated to release illegally copied files over the Winny network. That is because they might have been tracked with such IDs, where ID systems have been introduced into e-mail networks or other Internet services. Although such devices cannot completely stop illegal file transactions, they could be the next best things.

Dr. Kaneko seemed to have known the relations between these technology components and illegal file sharing judging from his message[10] on the Winny BBS. If he had introduced such components, however, Winny would not have been as widely used. He seemed to place diffusion of Winny ahead of preventing its illegal use.

10.3 Road to Winny *Criminal* Case

The Winny case traces a trajectory with the following chronology.[11]

The first arrest in the world for copyright infringement by file sharing software occurred in Japan on November 28, 2001. The software used by the arrested men was WinMX, which was then the most popular file sharing software. Its developer was not arrested. Afterward, the use of WinMX slowed down.

Mr. 47 (Dr. Kaneko's nickname on the BBS) appeared for the first time on April 1, 2002, in a thread of 2 Channel, i.e., the most popular open Internet BBS in Japan, and he declared the development of another file sharing software. The title of the thread was "What will the successor to *WinMX* be called?"

The name of the software he was developing was determined to be Winny on the 5th of the same month, which followed a suggestion by his supporters on the thread. Winny (WinNY) means the successor of WinMX (M $\rightarrow$ N, X $\rightarrow$ Y).[12]

The beta version of Winny was published on March 6, 2002. Then, supporters started testing the software. The first official version of Winny was published at the end of 2002. Upgraded versions of Winny were continuously published until the police began searching Dr. Kaneko's home and office.

Warnings on illegal file sharing first appeared in README attached to Winny on February 2, 2003. That is because Dr. Kaneko started a new way of distributing

---

[9] Such an ID function must not be implemented in an alternative system of Freenet. Even if it were implemented in the system for a verification experiment, it would not have caused substantial problems.

[10] Retrieved July 1, 2005, from http://Winny.info/2ch/2ch_log1.html (dead link).

[11] The chronology is based on the summary retrieved July 1, 2005, from http://www.nan.sakura.ne.jp/Winny/page/lib/history.htm (dead link).

[12] This episode strongly suggested that Dr. Kaneko recognized his software would be frequently used for illegal file sharing.

Winny. Winny had primarily been posted on his website before this, and such warnings were provided on it. He started to distribute updated versions of Winny from this new version through the Winny network. Thus, such independent distributions of README were required.

Development of the first generation of Winny was terminated on May 5, 2003. Beta tests of the second generation of Winny started simultaneously. Dr. Kaneko added an anonymous BBS function to the second generation.

The first PC virus targeting Winny users (i.e., an anti-Winny computer virus) appeared on August 8, 2003. The first anti-Winny virus incident was reported on March 31, 2004. Investigating information was ironically released on the Winny network from an affected PC owned by the Kyoto police. Similar incidents in public and private organizations continuously occurred.

Dr. Kaneko published a paper entitled "Digital content distribution system maintained by digital securities" on October 10, 2003.

Two people[13] who were alleged to have uploaded files illegally on the Winny network were arrested on November 28, 2003. Police simultaneously searched Dr. Kaneko's house and office. They seized items like his laptop and notebooks. As a result, the development of Winny ceased. However, this house search was officially said to be to collect evidence on copyright infringements by the two parties who had been arrested.

Dr. Kaneko was asked to go voluntarily to the police for questioning on May 10, 2004, and he was then arrested on charges of aiding and abetting copyright infringements. He received much media exposure, which reported his private and public life including his job as an assistant professor at The University of Tokyo.

10.4 Initial Responses by Society

Three typical initial responses by society to Dr. Kaneko's arrest were observed. These responses can be labeled as those by defenders, offenders, and meta-analysts.

The "Dr. Kaneko support group", which was formed three days after he had been arrested, is taken to be a typical defender.[14] Their emergency statement argued the following things: His arrest on charges of aiding and abetting copyright

---

[13] They were later convicted for copyright infringements.

[14] There could be observed another *famous* arrest of a software engineer in Okazaki "Librahack" case (http://astand.asahi.com/magazine/wrnational/special/2011012800004.html), where the engineer was arrested on suspicion of unlawful access to a library's server (DoS attack) was put under investigation for *weeks*. As the result of the prosecution's ignorance about technology, he was only suspended of prosecution, even though the case should have been just dropped because of insufficient evidence. Actually, he only used his software to access the server as frequently as once in a second for only a limited time, but these accesses "crashed" the server due to errors in the server program. As this case might represent, Japanese software engineering community was clearly skeptical about technical understandings of law enforcements.

infringements is unlawful because it was based on a stretched interpretation of the concept of aiding and abetting in criminal law. If developers of software are arrested on charges that its users exploit it illegally, most manufacturers and distributors of recording media, hardware, and software, which are not only used for legal copies but for illegal copies, could be arrested. That is irrational. If such an irrational arrest were accepted, many developers and manufacturers would cower at the risk of being arrested. This situation would be against the national interest.

In contrast, the Sankei group[15] news site, which considered him an offender, claimed that Dr. Kaneko's challenge to the copyright system was foolishly bold and obviously illegal. Their articles[16] described him and his acts as follows; "[He] was devoted to the development of unprecedented underground software." "A computer wizard competing head on against the copyright system was defeated." "[He] wrote messages on BBS like 'We cannot help redefining the concept of copyrights due to the appearance of anonymous file sharing software'." In sum, they insisted that a green computer whiz developed illegal software to challenge the copyright system as a result of his ignorance.

These two "inflated" responses were opposite opinions, but they shared four common features. First, neither of them drew little attention to "technological" aspects as the author did. They consequently missed why Winny was problematic within this particular context. In reality, Winny was not problematic just because it offered a file-sharing function. Second, offenders especially downplayed the legal details of the Winny case. Winny as software has been indisputably legal under the Copyrights Act. As will be explained later, the main point referred to by judicial courts was neither the legitimacy of Winny nor the legitimacy to develop it but whether Kaneko *intentionally* aided and abetted the infringement of copyrights.[17] That is, both sides only argued a fraction of his acts. Manufacturers and distributors actually do not need to recoil at the risk of being arrested, and Kaneko did not develop underground software. Third, both responses argued indirectly about what type of software development society should accept (and promote) on the basis of discussions on the Winny case. That is, defenders tended to claim that software developers should be defended for (potential) national interests even if there is minor damage in the process of its development and use. Meanwhile, offenders were likely to claim that software developers should be blamed for distributing software to be used for damaging existing legal systems even if the software brings important benefits to society, at least, in the future. Last, both discussed the Winny criminal case on the assumption that our copyright system was well organized and socially acceptable. For example, they did not mention that the system had become dysfunctional.

---

[15] Sankei group is a representative conservative media group.

[16] Retrieved July 1, 2005, from http://www.zakzak.co.jp/top/2004_05/t2004051119.html (dead link).

[17] Legal experts pointed out this central issue in a suit just after he had been arrested (cf. Sonoda, Asahi Newspaper, 14/05/2004).

Meta-analysts focused on the last feature. They attempted a bird's-eye analysis of the Winny criminal case. Their claim was a typical academic response. For example, Dr. Azuma,[18] a spirited cultural interpreter and academician, insisted as follows: It is necessary to formulate a new style of copyright controls and new billing systems for content when facing new technologies like Winny. Even if society blindly made such technologies illegal, such an order could not last long. Furthermore, it is nonsense to arrest software developers expanding technological frontiers. That is obviously against social and cultural interests in the long run. What we then need to discuss is problems with our current copyright system that is less attuned to innovations like the Internet and digitalization of content.

Copyright systems were historically introduced in response to the emergence of new technologies, and they have been changing with advances in ICT. In this sense, the claim of meta-analysts is hardly deniable, or only such discussions could provide a *final* solution to the matter. Still, as their arguments lacked analysis of the technological and legal aspects of the Winny case, they were nothing more than words on paper.

In view of those responses, the arrest of Dr. Kaneko aroused five main underlying issues: (1) Was his conduct legal? (2) Was his conduct socially acceptable? (3) What types of conduct by software engineers (or programmers) are legal? (4) What types of conduct by software engineers are socially acceptable? (5) What is a socially acceptable copyright system?

10.5 Defining *Problems*

If your private photographs are distributed over the Winny network,[19] how can you cope with this situation? As presented earlier, Winny is a P2P file sharing software to allow users to upload and download files over the Internet. Problems like this occur very often with or without Winny. However, unlike in cases of file sharing through the Web or the file transfer protocol (FTP), files uploaded through Winny automatically flow over its P2P network. To make matters worse, no one can stop this flow virtually, even in part.

Winny is likely to cause problems ironically because of its technological excellence. As was previously explained, it is very difficult to identify who has uploaded files using Winny. That is to say, the anonymity of uploaders is highly secure on its network. Consequently, Winny lowers the technical and psychological hurdles in illegal file sharing.

As a result, there are no measures to cope with these situations because people who uploaded personal pictures on the Winny network can hardly be found.

---

[18] Retrieved July 1, 2005, from http://www.hirokiazuma.com/blog (dead link).

[19] One of the most repugnant cases is the distribution of child pornography materials. In addition, there were known to be numerous disastrous incidents of private photographs being leaked.

Charges might have to be brought against the developer as its original cause may be attributed to the person who developed and distributed Winny. The developer, Dr. Kaneko, was actually arrested on charges of assistance in the infringement of copyrights without victims' complaints. However, this was not because he was alleged to have assisted someone to upload private photographs of victims.[20]

His first trial started in a District Court in September 2004 and his last trial ended in the Supreme Court in December 2011. People first expected to find an answer to their question (i.e., was his conduct legal or illegal because programmers are allowed or are not allowed to do things based on social values) through these trials with their *reasonable* belief that the judicial decision would reflect the social value of Winny and software in general. Therefore, this question addressed a problem to be solved concerning Winny.

In parallel with discussions on the above issue, courts faced another but somehow similar issue. That is to say, as there was no clause related to the assistance of copyright infringements, especially those by using technology in the Copyright Act, courts had to rule on whether Dr. Kaneko's actions were legal based on their interpretation of aiding and abetting in the Penal Code. Although he could be said to have assisted copyright infringements by using technology he developed, what he actually did on a superficial level was nothing more than releasing his versatile software on the Internet. On top of that, the court had to construct a rationale from this legal boundary of aiding and abetting, as there had been no direct judicial precedent. Thus, it was a matter of course that the final decision would serve as an important precedent. Thus, this issue was another problem to be solved concerning Winny.

The third and last problem was how to cope with illegal file sharing not only by using Winny but also by other means on the Internet.

It has been very hard to estimate the economic damage of copyright infringements by file sharing software. However, 26 million US dollars in copyright royalties were offered in a compromise settlement in Napster versus numerous record companies, for example [21] Although the Japanese music market is much smaller than that in the US, it is the second largest market in the world.[22] Thus, it is quite reasonable to estimate the economic damage caused by illegal file sharing in Japan has been huge. If illegal sharing of video content had gained momentum as the bandwidth of the Internet widened, there might have been much larger economic damage for digital content distributors throughout Japan.

---

[20] There might indeed been something wrong with the fact that he only stood trial in a criminal court over aiding and abetting of copyright infringements. However, there was no way for him to go to court without victims' accusations. It was natural for victims of private photographic outflows not to bring about action for damages considering the possibility of secondary or tertiary harm. Nevertheless, these outflows posed a serious social problem.

[21] Retrieved August 5, 2014, from http://www.jiten.com/dicmi/docs/n/8079.htm.

[22] Retrieved August 5, 2014, from http://ifpi.org/news/music-subscription-revenues-help-drive-growth-in-most-major-markets.

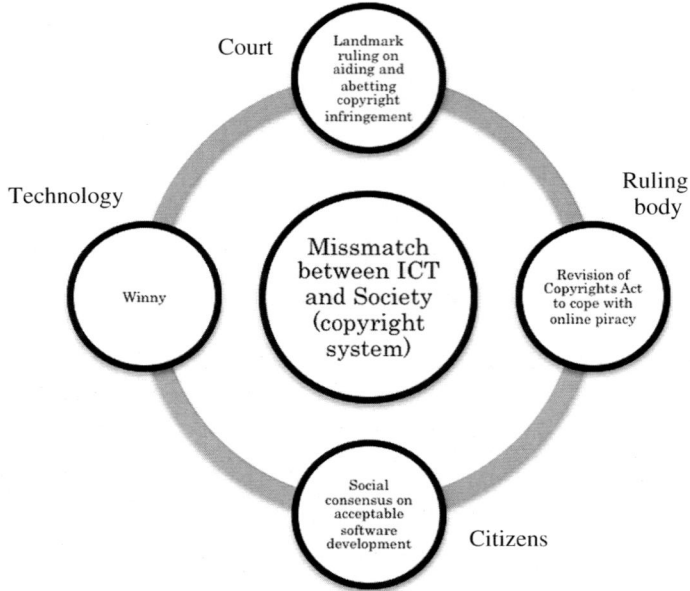

Fig. 10.1 Problems to be solved concerning *mismatch* between ICT and society

As was previously explained, discussions on illegal file sharing tended to focus on copyright infringements because of its economic impact and volume of stakeholders' voices. In this sense, the question that was asked at the beginning of this section would better be rewritten as "If your own CD data were distributed over the Winny network, how would you cope with this situation?" Meanwhile, we should never forget that the whole story was not to discuss Winny from the viewpoint of copyrights.[23]

In any case, people needed consensus building on how our society would cope with software like Winny (i.e., the 2nd problem), and our legislative body (virtually equal to the copyright protecting agency) had to revise the Copyright Act in line with the current conditions of technology and society (i.e., the 3rd problem).

The above situation is summarized in Fig. 10.1. These three problems are strongly linked to one another. Thus, if society had tried to solve them collectively and coordinately, we might have developed better solutions.

---

[23] For example, Ootani (2004b) discussed Winny from the viewpoint of "privacy". While, Lessig, who detailed privacy problems of the Internet in a chapter of his milestone book, Code and Other Laws of Cyberspace, Basic Books, New York (1999), pointed out that "Many people had understood privacy problems on the Internet." [Retrieved July 1, 2005, from http://www.asahi.com/tech/lessig/02.html (dead link)].

10.6 Winny Trials

Dr. Kaneko had been on trial on suspicion of aiding and abetting copyright infringements. The underlying issues in his case seemed to be exhausted on the first day of the trial at the District Court. There was only one important issue between the prosecution and the defense.

His actions under the US copyright system and laws would be referred to a court as to whether they were a contributory infringement.[24] Judging from the Betamax decision by the Supreme Court in 1984 (Sony Corp. of America vs. Universal City Studios, Inc., 464 U.S. 417), the focus could have been on whether Winny had substantially been used for non-infringing use, if the trial had been held in the USA. It would consequently have been judged in a US court according to how it was to be used in society (i.e., use value) as to whether Winny was legal or not.

In contrast, court decisions in Japan are not based on the way software is used, but on the intentions of its developer (i.e., criminal intent). Thus, such decisions that are irrelevant to the use value miss the point of questions like what type of software development society should accept.

In any case, our Copyright Act adopts a system that is different from that in the US. Thus, whether someone aided and abetted (i.e., assisted) copyright infringement is referred to a judicial court within the framework of criminal law. Requisites for enactment of *aiding and abetting* under the criminal law are: (1) the existence of principal(s), (2) assisting illegal acts of the principal(s), and (3) recognition of those assisting those illegal acts.

Only recognition of those assisting the principals' illegal acts became the main focused in the Winny case. This was because two Winny users had already been arrested for copyright infringement so that (1) and (2) were incontestable.

Each court defined this notion of *recognition of assisting copyright infringement*, which had been a focus of trials, as follows. The decisions of courts[25] were made according to the definitions and their findings.

The ruling in the trial by the District Court stated: "It is determined by the usage situation of technology in question and its social recognition as well as the supplier's subjective recognition about its distribution whether the action of supplying the technology per se is regarded illegal as aiding and abetting." And, it pointed out, "Although Dr. Kaneko obviously recognized and admitted that many users of Winny would use the software for illegal file sharing to infringe copyrights, he had continued to release and provide it to the public." It added, "He facilitated the two principals to share copyrighted files materially by releasing Winny 2 (i.e., Winny ver. 2.x) and psychologically by implementing the function of anonymity in

---

[24] Contributory infringement involves liability to promote infringement.

[25] The following quotations concerning courts' rulings are translated from the decisions of the Supreme Court and the High Court, which also contain the summaries of the decisions of the District Court.

Winny." Thus, it concluded that he was guilty of aiding and abetting copyright infringements.

The ruling in the succeeding trial by the High Court criticized the definition of aiding and abetting in the ruling of the District Court as follows: "The definition was short on specifics. Thus, the definition would make it possible to recognize even a provider of value-neutral technology as guilty from the fact that he or she was aware of the possibility that someone could use the technology for the wrong purpose."

The High Court overruled the original decision by indicating: "Judging from the spirit of the Copyright Act, foreign judicial precedents, and the viewpoint of harmonization between copyright protection and technology development, provisions of value-neutral technology to the public should not be recognized as aiding and abetting in principle. The provider could not be recognized as aiding and abetting, just because he or she was aware of the possibility that someone could misuse the technology."

Based on this discussion, it concluded, "The fact that a value-free software provider recognized the possibility that someone could use it illegally is not sufficient for assessing that the distribution of the software over the Internet facilitated the principal to share copyrighted files illegally. In addition to the fact, only if the provider encouraged specified or unspecified Internet users to use the software only or mostly for illegal file sharing, the provider should be judged as aiding and abetting copyright infringement." Then, by indicating the fact that Dr. Kaneko reminded Winny users not to use it for illegal file sharing in its README file, the court did not recognize that he had encouraged specified or unspecified Internet users to use Winny only or mostly for illegal file sharing. Consequently, he was acquitted of aiding and abetting copyright infringement.

The ruling in the last trial by the Supreme Court stated: "The prerequisite for establishment of aiding and abetting is both the existence of concrete status of infringing software usage and its provider's recognition of the status." On that basis, the court defined two situations for its establishment. (A) "While a provider recognized that concrete infringing usage was about to happen, the provider still continued to release the software. As a result, the software was used for copyright infringement." Or, (B) "The following conditions were right: (1) it was highly probable that an unexceptional number of people who obtained possession of the software in question would use it for copyright infringement judging from the context of the features of the software, objective status of its usage, a way of providing it, and so forth, (2) the provider of the software recognized this probability and still released or provided it, and (3) someone (i.e., the principal) used it for copyright infringement." On this basis, the Supreme Court suggested, "It is a matter of course that situation (A) is not applicable, because Dr. Kaneko was not acquainted with the two principals." Meanwhile, the court admitted, "It was highly probable that an unexceptional number of people who obtained possession of Winny would use it for copyright infringement, and he recognized this probability but still released or provided it" from an *objective* point of view. Then, the court said, "From his *subjective* viewpoint, it is admissible that he recognized some

Winny users infringed others' copyrights and that the number of such users was increasing. However, there was not enough evidence to admit that he recognized it was highly probable that an unexceptional number of Winny users infringed others' copyrights." Thus, the court denied the establishment of situation (B), and it returned a verdict of not guilty. In other words, the Supreme Court admitted that Dr. Kaneko had assisted copyright infringements from objective points of view. As there was not enough evidence to prove that he had been aware that he could have assisted copyright infringements due to his subjectivity, the court acquitted him.

The Supreme Court presented the reason why such complicated logic was adopted in the trials as follows: "As the District Court and High Court called Winny value-free software, it could be used for legal purposes as well as illegal purposes like copyright infringements. Thus, whether Winny is used for illegal purposes or other purposes is consistently left to the judgment of the individual." In this sense, it is extremely difficult to judge whether aiding and abetting copyright infringements can be established, based only on the fact that software used for copyright infringements was provided to the public.

Furthermore, as the Supreme Court intentionally chose the word of *called* in the above quote, it did not admit the existence of a priori value-freeness of technology, which STS perceives in a negative light. While, the rulings of the District Court and the High Court were based more or less on a different notion of value-freeness from that of STS.

The District Court said concerning the value-freeness of technology in its trial: "Winny is P2P file sharing software. As clearly described in the defender's testimony and statements, the software is *worthwhile and applicable in various fields* as a technological implementation of P2P requiring no central server. In that sense, regardless of his intention of development, the technology developed per se is value-free." In its ruling, although the technology of Winny was broken down at the level of element technologies, the High Court said: "No element technology of Winny is specialized for copyright infringements, thus Winny is value-free software, namely, software to facilitate its users to exchange information efficiently with secrecy as well as to infringe others' copyrights." In both cases, the possibility to use technology in question in various ways is regarded as evidence of value-freeness of technology.

In any case, as this judgment by the Supreme Court served as a precedent, without taking into consideration the social value of technology that should only be judged within each specified context [i.e., neglecting the value of technology that is socially constructed (Pinch and Bijker 1984)], the legality of development actions by engineers (or, at least, software engineers) can be judged according to a very limited perspective, i.e., based on the engineers' recognition of the current situation. As a result, although the series of trials finished, the problem that surfaced due to the arrest of Dr. Kaneko (i.e., What is socially acceptable software development?) remained unsolved. On top of that, the judicial precedent to prevent social consensus on acceptable software development from being used in court decisions was settled.

10.7 Historical Circumstances of Japanese Copyright and *Solutions* to Winny Problems

Behind the Winny criminal case were such factors shared by developed countries as rapid development and diffusion of ICT (e.g., the Internet, digital content, and personal computing) flaws in copyright systems that could not catch up with such rapid developments of ICT, and limitations of technological measures to prevent copyright infringements. Actually, Dr. Kaneko proposed a *solution* (a digital content distribution system maintained by digital securities) by considering these factors. The Supreme Court adapted this fact as (minor) evidence to prove his innocence.

In addition to this common background, the Japanese *advanced* system of copyrights deserves attention as a factor unique to Japan. First of all, copyright and related rights, especially the latter tend to be protected heavily in comparison with other countries. More to the point, old-established content industries have been protected more than sufficiently in Japan. For example, in terms of prices of the top 20 CD albums in the early 2000s, the US average was 55, and the UK, French, and German averages ranged between 65 and 70, when the Japanese average was set at 100.[26] In addition to these price gaps, there is another factor that protects content industries, namely, a resale price maintenance system. Thanks to the system, covered content like that in books, magazines, newspapers, and music media are exempt from antitrust laws.[27] Thus, industries cannot only place high price tags on such content but can also maintain their prices. As a result, Japanese consumers have been compelled to accept substantially higher prices for content than those in other developed countries.[28]

Under these circumstances, the Japanese copyright system is characterized by industry protection as is natural. Because of this characteristic and others, Japan is sometimes said to be an *advanced* country in terms of copyright protection. For example, Japanese lawmakers responded to the World Intellectual Property Organization (WIPO) Copyright Treaty (ratified in 1996) at a moment's notice and were the first in the world (1997) to introduce the "right of making transmittable" and the "public transmission right" into the Copyright Act. In comparison, the US introduced them into the Digital Millennium Copyright Act in 1998, and the EU into 2001/29/EC in 2002. Japan was the earliest country to follow these rights and started to provide "technical protection" and "protection of rights management

---

[26] Japan Fair Trade Commission (2004), a material distributed in its working group meeting.

[27] Digital distributions are excluded from the system. Movie and computer software are also excluded from the system.

[28] There is a "bunko" (Japanese style of paperbacks) system in Japan, where (mainly popular) books are published as paperbacks at very low prices years after they have been published in hard cover. Thus, we cannot say that the prices of books are higher than those in developed countries. However, although movie content (typically DVDs) is not exempt from antitrust laws, its prices, against expectations, are 20–50 % higher than those in other developed countries.

information" legally in 1999, in response to the WIPO treaty. Moreover, the irony is that Japan is the first country to arrest people for illegal transmission of copyrighted content.

People or groups who appreciate Japan for its *advanced* system of copyrights are possibly, in an extreme instance, members of copyright protecting bodies and industries that have earnings from related rights. In fact, some experts and scholars have strongly criticized the Japanese copyright system as being anachronistic. Why are there different assessments that are 180-degrees apart for the same system? The following are part of the reasons.

1. Copyright protection features alone are strengthened in the Japanese copyright system. Thus, these features are assessed as *advanced*.
2. Users' rights are uncertain and more limited than those in other countries. For example, the range of fair use is not defined in the Copyright Act, which only lists items like those in private use and quotes in its individual exceptional rules. It is symbolic that Japan has prohibited reproduction for private use concerning the "circumvention of technical protection" since its introduction. However, other countries did follow Japan later.
3. A strange clause (Preventive measures for music record reimport) was added to the Copyright Act in 2005. This clause prohibits third parties from importing music content media sold overseas by copyright and related rights holders under the Copyright Act. Thanks to the clause, disparities between domestic and foreign prices can be maintained.

Nevertheless, it has to be pointed that their copyright protection measures had been tolerant to illegal file downloading until 2012 from the standpoint of fairness. That is quite in contrast to the US, where a rush of claims for damages[29] inflicted by illegal file downloaders by the Recording Industry Association of America (RIAA) and record companies occurred in the mid-2000s. Concerning this, it is particularly worth noting that almost all cases of illegal file sharing have not to civil cases (e.g., claims for damages and injunction demands) but to criminal cases as was exemplified by the Winny case. Illegal reproduced copies of business application software, which often lead to damage suits, are exceptional though.

Although it is completely speculative, a few acceptable reasons for tolerance to illegal file downloading could be explained as follows. (1) Copyright protecting bodies assessed that it was sufficient to crack down on illegal file uploading. (2) They tried to avoid rank-and-file resentment (i.e., sever conflicts with citizens over copyright protection), which might have occurred if they controlled actions by *innocent* citizens' in downloading contents. Whatever the case may be, cracking down on illegal file uploading could not stop illegal file sharing Then, downloading copyrighted files became illegal without penalties in 2010 and with penalties in 2012.

---

[29] One of the most tragic outcomes would be Thomas-Rasset versus Capitol Records, 12-715 (Retrieved August 5, 2014, from http://www.supremecourt.gov/Search.aspx?FileName=/docketfiles/12-715.htm).

A brief history from the PC and Internet booms to the death of Dr. Kaneko is roughly compiled into a chronology in Table 10.1. It reveals a *strange coincidence*. That is to say, arrests including that of Dr. Kaneko were made as if keeping pace with the tightening of regulations for illegal file sharing. The tightest regulation measures were implemented almost as soon as he was acquitted.

The author has no intention to insist that this *coincidence* occurred for a purpose. Trials in courts and revisions of the Copyright Act are institutionally independent of each other. There is actually no way for them to directly interact. Nevertheless, things were consistently being shaken down. Thus, two controversial STS problems (i.e., Why, how, when, and to what extent does our society regulate illegal file sharing? What is socially acceptable software development?) were *solved* in a sense. Actually, although there were social debates on how to regulate illegal file sharing software mainly on mass media but not in the area of decision-making, conflicts in their main points did not become so obvious. That is, the problem of socially acceptable software development not only remained unsolved but was also avoided. Moreover, the problems were *solved*, while vested interests that had gradually been shrinking with the increase in digital distribution were protected.

Yet another Japonism that insists that such strange *problem solving* is a characteristic of Japanese society has little basis in fact. Instead it can be concluded that players do not solve problems directly but try to handle situations while avoiding

Table 10.1 Chronology of Winny and Japanese copyright system

	Upload	Download	Event
1995	Legal	Legal	Windows95 was released
1996			WIPO Copyright Treaty was adopted in response to advances in information technology
1997	Illegalized with penalty		Revision of the Copyright Act: *The first illegalization* of this sort among developed countries
2001			File sharing software, WinMX, was developed and released
			The first arrest in the world: Two WinMX users, who uploaded software and content, were arrested
2002			Winny was developed and released
2003			Two Winny users, who uploaded software and content to the Winny P2P network, were arrested
2004			Winny developer (Dr. Kaneko) was arrested
			Two Winny users were found guilty
2006			Dr. Kaneko was convicted in District Court
2009			Dr. Kaneko was acquitted in High Court
2010		Illegalized without penalty	Revision of the Copyright Act
2011			Supreme Court absolved Dr. Kaneko
2012		Illegalized with penalty	Revision of the Copyright Act
2013			Dr. Kaneko passed away

serious conflicts among stakeholders. As a result, problems became less problematic as to what extent they could be shelved.

The content industry has avoided conflicts with consumers by turning down civil suits presumably, which could be the shortest road for problem solving. Copyright protecting bodies did not select quick and comprehensive measures to regulate file sharing but strengthened their regulations step by step so citizens could gradually adjust to their regulations. Minimal legal orders concerning file sharing were maintained in this process by the presence of a *scapegoat*.

Prosecutors and defense counsels in the process of the court trials respected visionary issues like value-free technology and developers' intentions; thus, the courts could avoid complicated and irresolvable problems. Even so, the Supreme Court issued a moderate ruling for software engineers, and it defined sorts of minimum guidelines for software engineers.

What problems did these results incur? It is needless to say that the life of a software engineer was sacrificed. Not only that, many victims, who had their privacies invaded through the Winny P2P network, remained discontented, indirectly because of the engineer's omissions. It is even worth mentioning that their difficult experiences as tragedies have been virtually neglected by society. They were just treated as amusing anecdotes. Nevertheless, Dr. Kaneko was not held legally responsible, even though he had been excessively punished by society.

To make matter worse, the row over Winny has ended and our software engineering community does not seem to have learned substantial lessons on what to do with themselves from this series of tragedies[30] and our society still does not have any clue about what socially acceptable software engineering is.

10.8 Conclusions

Main questions asked in our society in the Winny criminal case could be described as follows: (1) Is Winny legal? (2) Is Winny socially acceptable? (3) What type of software development is legal? (4) What type of software development is socially acceptable? (5) How can we build *reasonable* social systems where copyright protection and ICT innovations are compatible?

Our courts can basically only answer the 1st question due to legal constraints. If the Supreme Court should issue a landmark ruling as originally expected, it would be an answer to the 3rd question. Instead of that, the court issued its ruling that forces us to judge the legality of software development not by social values but by the developer's subjective recognition of various contexts. Subsequently, how will our society lead in socially acceptable software development? We merely have to wait for legal revisions if there is a huge change in our attitudes concerning the 4th

---

[30] See the memorial address at the beginning. It only focused on how society should treat software, software engineers, and innovations.

question, or if there are overly destructive innovations in ICT for our legal system to control.

The question "What type of software engineers' conduct (software development) is legal?" could be a difficult question for courts to answer, if there are no laws directly linked to it. Probably, "anything goes" would be a universally acceptable (or inevitable) answer.[31] Nevertheless, without directly applicable laws and regulations, Japanese courts are sometimes expected to regulate software engineers' conduct through their judgments, as was indicated in the Winny criminal case. In preparation for such situations, it might be a provision to introduce US style legal principles, where the *value* of technology is not judged by its purposes of development but by its use values. This is because this introduction makes it possible to indirectly incorporate social values into legal decisions.

It can be another comprehensive solution to secure a direct channel for *social values* to be reflected in legal decisions, which could have been achieved by that Supreme Court decision. The Supreme Court admitted in its actual decision that Dr. Kaneko had assisted copyright infringements from objective points of view. Because there was not enough evidence to prove that he had been aware that he would have assisted copyright infringements due to his subjectivity, the court acquitted him. If the court decision had not been based on *his subjectivity* but on the objective evaluation of whether he should have recognized that he would assist in copyright infringements within his context (e.g., he was a software engineer with a doctorate in the field and/or an assistant professor of software engineering), the decision could have reflected social values, and a code of ethics for software engineers to some extent. Thus, the technological trajectory of the field might have reflected such values as well. The Supreme Court could have acquitted him even with this changing basis of argument.

Unfortunately, the link between problems bridged by Winny was broken into pieces as the key decision makers closed themselves off from other problem areas and our society gradually shifted its attention away from Winny. As a result, the problems as a whole went by the wayside although each problem obtained its local optimum (Fig. 10.2). Indeed, although our society did not have any clues about what socially acceptable software engineering is, it returned to a stable condition. This *solution* may remind us *translation* in Actor Network Theory (Callon 1986).

We can suggest problems (or dysfunctions) of political systems, especially those between legislation and administration as the reason why our society did not seek a global optimal solution. Unfortunately, legislative system in Japan does not function properly because it has a distorted power structure.[32] That is to say, both parliaments and bureaucrats can make new laws and reform old laws in a realistic

---

[31] There are currently exceptions like "circumventions of technical protection" in many countries.

[32] Nonaka (2014) described this situation as an externally powerful parliament that has become a dead letter.

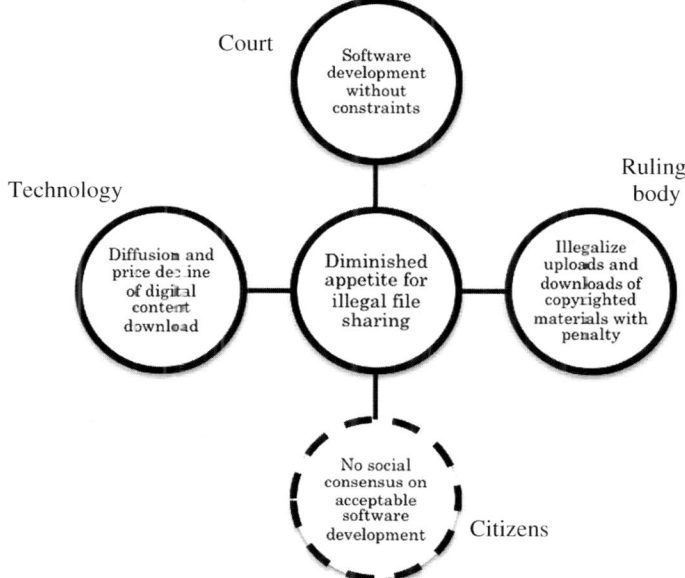

Fig. 10.2 Tentative *solutions* to mismatch between ICT and society

sense, as the case now stands.[33] As electorates generally lack interest in issues like copyright problems, parliamentary members avoid becoming closely involved with the problems. Consequently, this area has become the exclusive province of bureaucrats. To make matters worse, as bureaucrats tend to follow existing frameworks and to develop cooperative relations with relevant industries, we cannot expect changes to the current industry-friendly framework of the copyright system. Therefore, a malfunctioning copyright system continues to be used in Japan, and political attention, which was attracted once by the Winny criminal case, has turned out to be forgotten and neglected.

Safety questions about nuclear power plants had not clearly emerged (i.e., they had not become a *real* political agenda nationwide[34]) until the Great East Japan Earthquake as the agenda was supported by a similar political structure. Because most citizens had not taken safety questions seriously to the extent that it became the main agenda of our national elections, a *nuclear village*, which is a complex bureaucratic system with our power industry (and relevant academics, politicians), was easily maintained in an unrecognizable manner. That could have been a remote but essential cause of the accident.

---

[33] Only parliamentary members can theoretically make and reform laws by statute. However, there are open loopholes.

[34] It could be the case that they would become a real political agenda at the local government level. For example, there is a well-known case where a power company conceded plans to build nuclear plants in "Maki Machi" (Maki Town).

References

Callon, M. (1986). Some elements of a sociology of translation: domestication of the scallops and the fishermen of St. Brieuc Bay. In J. Law (Ed.), *Power, action and belief: A new sociology of knowledge*. London: Routledge & Kegan Paul.

Lessig, L. (1999). *CODE and other laws of cyberspace*. New York: Basic Books.

Nonaka, N. (2014). Toppagata-seiji ni Morosa (Pitfall of "break through" politics), *Nikkei Newspaper* (July 28, 2013).

Ootani, T. (2004a). Winny kaihatsusha no taiho ha nani wo imisurunoka (What does the arrest of Winny developer mean?). *Kagaku, 74*(8), 935–938.

Ootani, T. (2004b). P2P to joho-rinri (P2P and information ethics). *In: Proceedings of the 3rd Annual Meeting of Japanese Society for Science and Technology Studies* (pp. 19–22).

Pinch, T., & Bijker, W. (1984). The social construction of facts and artefacts: Or how the sociology of science and the sociology of technology might benefit each other. *Social Studies of Science, 14*, 399–441.

Shirabe, M. (2005). Sai-sentan gijyutu to hou: Winny jiken kara (Advanced technology and law; the case of Winny). In Y. Fujigaki (Ed.), *Kagaku-gijyutu-shakai-ron no gihou (Case analysis and theoretical concepts for science and technology studies)*. Tokyo: University of Tokyo Press.

Tanahashi, H. (1997). Computer network ni-okeru horitsu-mondai to genjyo-deno taisaku (Legal problems and measures concerning computer networks) (1)–(7). NBL No. 615–625.

The High Court of Osaka Japan. (2009). *Chosaku-ken-hou ihan houjo hikoku jiken (Case of aiding and abetting of copyright infringements)*. Dated October 8, 2009. Its copy was retrieved August 5, 2014, from http://danblog.cocolog-nifty.com/index/2009/12/winny-5bc6.html.

The Supreme Court of Japan. (2011). *Chosaku-ken-hou ihan houjo hikoku jiken (Case of aiding and abetting of copyright infringements)*. Dated December 19, 2011. Retrieved August 5, 2014, from http://www.courts.go.jp/app/files/hanrei_jp/846/081846_hanrei.pdf.

Yamamoto, T. (2004). *America chosaku-ken-hou no kiso-chishiki (Basic Knowledge about Copyright law in the US)*. Tokyo: Ota Publishing.

Index

A
Accountability, 47, 49, 52
Acquired Immunodeficiency Syndrome (AIDS), 3, 38, 161, 195, 197, 202–206, 209, 215
Administrative litigation, 167, 170–172, 176, 181, 183, 185, 186, 188, 190, 192
Administrative litigation, substantial judgement, 185, 188
Advocation, 31, 32, 36, 39, 49, 50, 90, 101, 214
Agenda
 building, 28, 34, 44, 48, 51
 setting, 30, 31, 32, 44
Aiding and abetting, 219–221, 224, 227, 229–231
Ambiguity, 47, 81

B
Backtracking, 134, 135
Beck, U., 14
Beyond assumption, 13, 14, 16
Bijker, W., 15, 16, 231

C
Cadmium poisoning, 141, 142, 148, 155
Callon, M., 236
Center for Desease Control (CDC), 198, 203, 205, 206
Chernobyl
 accident, 14, 43, 49, 63, 66, 74
Committee on the Biological Effects of Ionizing Radiation (BEIR), 61
Communication disaster, 1, 7, 8, 16, 17, 19
Consensus conference, 95, 99
Construction permit, 169–172, 177
Copyright
 act, 223, 227–230, 233, 234
 infringement, 3, 219, 220, 223–225, 227, 229–232, 236
 system, 221, 225, 226, 229, 233, 237
Counteraction, 131, 133
Cryo, 201, 203, 204, 211, 214–216

D
Decision-making, 2, 17, 19, 21, 82, 88, 89, 97, 98, 111, 118, 168, 234
Deficit model, 33, 62, 64
Deliberation, 16, 48, 57, 58, 81, 94, 95, 98, 99, 105, 118, 119, 192
Deliberative Polling (DP), 19, 21, 87–89, 92, 93, 96, 98, 99, 103, 109, 112, 117
Democracy
 deficits, 79, 83
Democracy, democratic, 57, 59, 80
Democratic Party of Japan (DPJ), 76, 90, 93, 119
Demonstration, 118, 119
Desforges, J. F., 203
Discretion
 administrative, 2, 181, 185, 191
 expers', 185, 187–189, 193
Distrust, 17, 28, 59, 63, 113, 119
Dryzek, J., 95

E
Environmental regeneration, 141, 161, 162
Expert
 knowledge, 181, 184, 186
 witness, 185, 188

Expertise
  best available, 180

F
Fast-breeder rector, 2, 89, 167, 174
Felt, U., xi, 7, 97, 118
File sharing software, 219, 221, 226
Fishkin, J., 93, 94
Four major pollution-related lawsuits, 142, 151

G
Gallo, R., 206
Goyo-gakusha, 104
Green Cross Corporation, 199
Gunji, A., 202, 209

H
Hagino, 146, 147, 154
Hashimoto, 149, 150, 156
Heat-treated blood products, 202, 206, 208, 215, 216
High Court's decision, 171, 176, 179, 185, 186, 188, 191
Human Immunodeficiency Virus (HIV), 2, 3, 195, 197–199, 202, 211, 216

I
IAEA, 18
ICT, 228, 232, 236, 237
Impact, 37, 40, 74, 84, 96, 98, 111, 118, 173, 228
Indifference to time, 137
Infallibility, 33
Information disclosure, 17, 20, 21
Innovation, 33, 96, 220, 236
Integration, 15, 63
International Commission on Radiological Protection (ICRP), 41, 61, 64, 71
International Risk Governance Council (IRGC), 81
Invalidity confirmation, 172, 174, 175, 185
Itai-itai disease, 2, 11, 141–143, 146–151, 153–155, 159, 161, 162
Itai-itai disease residents' association, 151, 154, 158

J
Japan Science and Technology Agency (JST), 35
Jasanoff, S., xi, 168
Journalism, 30, 32, 112
Judgement
  administrative, 184, 187
  engineering, 167, 178, 181, 187, 193
  substantial, 185, 186, 187, 189, 192
Judicial decision, 153, 168, 227
Judicial system, 168, 169, 183, 185

K
Kainuma, H., 8
Kamioka mine, 141–143
Kamioka mine, 141–145, 147, 148, 150, 153, 154, 159
Kan, Naoto, 91
Kaneko, I., 219
Kisha-club, 30, 33, 34
Kobayashi, J., 34, 148
Kobayashi, T., 95

L
Lay experts, 51, 161
Legal
  system, 225, 236
Legitimacy, legitimate, 7, 13, 57–60, 71–73, 81, 113, 117, 131, 182, 185, 186, 190, 225
Linear Non-threshold (LNT), 61, 63, 66, 75
Low-dose radiation, 59, 60, 62, 65, 67, 73, 75, 81
Low-probability events, 172, 174, 176, 188, 193

M
Makino, J., 13, 50
Man-machine interface, 182
Matsunami, 142, 147, 149, 153, 154
Medical Care Standard of the Time ((MCST).), 195, 216
Meltdown accident, 173, 175
Minamata disease, 2, 83, 126–129, 131, 135, 151, 159, 160

Index 241

Mini-publics, 94, 96–98, 102
Monju, 2, 11, 103, 168–170, 172, 173, 177, 181, 185, 190

N
National discussion, 87, 88, 91, 101, 112, 115, 116
National Hemophilia Foundation (NHF), 203
Netizen, 32, 48
New technology(ies), 3, 10
New theory of negligence, 214–216
Non-heat-treated blood products, 195, 196, 199, 201, 203, 209, 211, 214, 216
Nuclear energy, 7–9, 13, 87, 90, 91, 101, 104, 106, 107, 114, 117, 120, 192
Nuclear Reactor Regulation Law, 176, 181, 182, 184, 191
Nuclear Safety Commission (NSC), 176–180, 191

O
Organic mercury, 126–128, 130, 133, 135

P
Paternalism, paternalistic, 21, 60
Pluthermal, 168
Policy
 energy, 41, 88, 90, 99, 102, 104, 168, 174
 nuclear, 87, 89, 90, 114, 119
Pollution control agreement, 2, 154, 160
Precaution, 51
Precautionary
 approach, 50, 80, 83
 attitude, 75
 governance, 83
 principle, 2, 3, 162
Procedural examination, 186, 188, 189
Public
 engagement, 3, 10, 21, 34
 participation, 2, 88, 89, 95, 97, 112, 115, 118, 120, 143, 162
 sphere, 2, 11, 48

R
Radiobrain (houshu-no), 75
Radiophobia, 66, 75
Reconstruction agency, 76, 78
Research Institute of Science and Technology for Society (RISTEX), 35
Resonance, 136, 138, 139
Responsibility, 17, 22, 47, 59, 61, 72, 129, 151, 168, 201, 216
Rhetoric, rhetorical, 57, 61, 73, 80

Right
 to avoid radiation, 74, 79
Risk
 communication, 11, 49, 57, 59, 78, 82
 comparison, 67, 68
 discourse, 58
 governance, 57, 58, 81
 perception, 68

S
Safety
 absolute, 177, 188
 relative, 188, 189, 193
 review, 173–175, 177–180, 184, 191
Salient value, 17, 18, 23
SAve Fukushima Children Lawyers' Network (SAFLAN), 77
Science
 function in society, 43, 125, 130
 journalism, 32
Science Council of Japan (SCJ), 17, 20, 62, 64
Science Media Centre, 1, 33, 35
Scienceplanner, 66
Scienceplanation, 57, 60, 62, 67, 71, 81
Second Minamata disease, 142, 151
Segregation
 between fields, 16
 between sites, 7, 10, 11, 15, 18, 19, 23
Severe accident, 177
Shigematsu, I., 149, 155, 161
Silent majority, 119, 120
Social
 consensus, 219, 231
 value, 227, 231, 236
Socio-scientific, 28, 30, 48, 51
Sodium leakage, 173, 178, 180
Stakeholder, 2, 43, 79, 81, 88, 97, 235
Steam generator tubes, 173
Stepwise regulation, 167, 177, 182, 191
Supreme Court's decision, 171, 181, 183, 186
Supreme Court's decision, Ikata NPP, 183–185

T
Tacit knowledge, 178
Takagi, H., 181, 188
Taketani, M., 9
Tanaka, Snozo, 147
Timeless, 137
Tokyo Electric Power Company (TEPCO), 12, 14, 31, 58, 59, 70, 73
Trust
 social, 18
Type I and Type II errors, 156

U
Ui, J., 131
Uncertainty
  of science, 62
Unexpected, 7, 8, 13, 139, 179
United Nations Scientific Committee on the Effects of Atomic Radiation (UNSCEAR), 63, 65, 66
Unknown risk, 180

V
Value
  free, 230, 231, 235
  freeness, 231
  neutral, 230

Victims Support Act, 59, 73, 74, 76, 77, 80
Voluntary evacuees, 75, 77

W
WIPO, 232
Working Group on Risk Management of Low-dose Radiation Exposure, 65
Wynne, B., 33, 62

Y
Yasui, I., 179
Yokka-ichi Pollution, 142
Yoshioka, H., 9, 12
Yoshioka, K., 148

Printed by Books on Demand, Germany